'Dr Thomas successfully weaves together current knowledge and research from the fields of sleep and brain injury with a vast scope that ranges from the relationship of the neuroanatomical and neuro-chemical foundations through to the lived experience of the individual. The book expertly constructs a deep appreciation of the complexity and interrelatedness of the extensive ecosystem that is sleep and waking as it structures and encompasses all aspects of life. Based on the intricacies of the above, a complete sleep management approach is offered to promote healthy sleep and waking, which makes clear how things might be improved for those who have experienced a brain injury. It is a must read for anyone working in this area.'

—Dr Susan Hooper, BSc (Hons), Cert. Ed,
MSc, D Couns Psych, Cert. Rehab
Chartered Psychologist, Registered
Counselling Psychologist, HCPC

'An extraordinary work, I have learned so much. A few items that stand out in particular include: the disturbing persistence of pro-inflammatory cytokines (and potential risk of neurodegenerative disease) for years post brain injury, the impact of melatonin and sleep deprivation on bone mineral density, the impact of reduced or excess REM on depressive symptoms, including suicidal ideation post TBI, the unsur-prisingly complex and beneficial role of Vitamin D, the microbiome and vagus nerve stimulation: all areas of research in inflammatory arthritis. This book offers a thorough review of the neuroscience of sleep and brain injury, together with novel evidence-based interventions to optimise outcomes.'

—Dr Jo MacGowan, BSc (Hons), MBBS, FRCP,
Consultant Rheumatologist

'This astonishingly well researched book integrates current knowledge from sleep theory and practice and the wider neurosciences to provide a comprehensive framework for understanding sleep disorders and brain injury, and their interaction. It will be an invaluable resource to guide all neuro-rehabilitation clinicians in assessing and treating such problems.'

—Dr Lesley Stewart, MA, MPhil,
Consultant Clinical Neuropsychologist, Chartered Psychologist, HCPC,
Full Practitioner Member Division of Neuropsychology

Sleep and Brain Injury

This ground-breaking book binds together a contemporary understanding of sleep and brain injury, pairing empirical understanding through clinical practice with extensive up-to-date research, to provide a deeply considered approach to these overlapping topics. Firstly, the author discusses the neuroanatomy and architecture of sleep, including the need for sleep, definitions of good sleep, and what can go wrong with sleep. The focus then moves to the neuroanatomical damage and dysfunction from brain injury, and the resultant functional effects. The author then adroitly fuses the two streams of thought together, focusing on the neurobiological, neurochemical, and functional aspects of both sleep and brain injury to offer new insights as to how they interrelate.

The book then looks towards the applied aspects of treatment and rehabilitation, bringing further thoughts of how, because of this new understanding, we can potentially offer novel treatments for brain injury recovery and sleep problems. In this final practical part, four sleep foundations are given, necessary to optimize the three most common sleep problems and their treatments after brain injury.

This new approach highlights how sleep can affect the specific functional effects of brain injury and how brain injury can exacerbate some of the specific functional effects of sleep problems, thus having the potential to transform the field of neurorehabilitation. It is essential reading for professionals working with brain injury and postgraduate students in clinical neuropsychology.

Crawford M. Thomas has been in healthcare and academe for over 30 years. He has designed, implemented, and led inpatient and community acquired brain injury neurorehabilitation services. A skilled Consultant Clinical Neuropsychologist and Neuroscientist, he is the clinical lead of one of the largest community acquired brain injury teams in the UK, working for Cheshire and Wirral Partnership NHS Foundation Trust.

Sleep and Brain Injury

Crawford M. Thomas

Routledge
Taylor & Francis Group

LONDON AND NEW YORK

First published 2022
by Routledge
2 Park Square, Milton Park, Abingdon, Oxon OX14 4RN

and by Routledge
605 Third Avenue, New York, NY 10158

Routledge is an imprint of the Taylor & Francis Group, an informa business

© 2022 Crawford M. Thomas

British Library Cataloguing-in-Publication Data
A catalogue record for this book is available from the British Library

Library of Congress Cataloging-in-Publication Data
A catalog record has been requested for this book

ISBN: 978-0-367-18896-2 (hbk)
ISBN: 978-0-367-18899-3 (pbk)
ISBN: 978-0-429-19906-6 (ebk)

DOI: 10.4324/9780429199066

Typeset in Times New Roman
by MPS Limited, Dehradun

This book binds together a contemporary understanding of sleep and brain injury. By pairing the author's empirical knowledge, gained over a quarter of a century in clinical practice, with extensive up-to-date research it sets the stage for a thoughtful and deeply considered approach to these overlapping topics. The author discusses the neuroanatomy, as well as the architecture of sleep. The subsequent analysis includes the need for sleep, definitions of good sleep, what can go wrong with sleep, and the resultant functional effects.

The focus then moves to brain injury. What is known about the neuroanatomical damage following common forms of brain injury and some of the resultant functional effects. The author then adroitly fuses the preceding streams of the book together, particularly the neurobiological, neurochemical, and functional aspects of each.

The conclusion brings further thoughts of how, as a consequence of this new understanding, we can potentially offer novel treatments for brain injury recovery and sleep problems. In this final practical section, four sleep foundations are given, necessary to optimize the three most common sleep problems and their treatments after brain injury.

This new approach is designed to help sleep and brain injury recovery together. Overall, its central concern is to allow sleep to heal. The book is aimed at all those working with people who have had any form of brain injury.

Contents

Preface

I feel that on each occasion I attempt to write this, I get lost in a long list of names, as expressing my gratitude to all those who have helped me during the course of my career, and this preface specifically would have constituted a book in itself. Suffice to say I appreciate all those unnamed, as well as named; they know who they are. So it is to those needing special mention: Lucy, Saloni, Alison, and all at Routledge for their patience and such polite prompting emails have been marvellous. My family both current and of origin, all of whom have provided such patience in this mammoth undertaking. It struck me on more than one occasion how amazing it is that anyone can write more than one book in their life. Having lived like a hermit for a few years of writing, I will emerge again through their love.

Finally, I would like to thank all the wonderful people in the team I work with in Cheshire and Wirral, especially Megan and Pip for their kindly organization of this disorganized soul. Above all, although she will demure, I want to say a very special honour goes to Beth. She is exactly the kind of manager the NHS (and Independent Health Sector) needs but has so few; exceptional frontal lobes and a true focus on the patient. Without her astonishing ability to have allowed me to carve out the time, I would never have achieved starting, let alone finishing this book.

Above all my patients, who over the years have taught me so much. I am ever astonished how so many of you manage to keep going in the eye of such a storm that is brain injury. I will continue to fight beside you as best I can.

Dr Crawford M. Thomas
BSc (Hons), D Clin Psych, Cert Psych, Dip Neuro, HCPC, Full Practitioner
Member of the Division of Neuropsychology, MBPsS

Foreword

Crawford and I became colleagues in 2005 whilst working at a tertiary inpatient neurorehabilitation centre providing specialist rehabilitation to patients with a brain injury and highly complex rehabilitation needs from across the UK. During the seven years that we worked together, I witnessed his dedication to patients and passion for brain injury rehabilitation, but besides clinical work there was always the willingness to question current clinical practice and improve outcomes for this patient group. It is no surprise therefore that he has written this very comprehensive book which takes us through sleep and brain injury and how the two link in, interplay, and profoundly affect each other.

He starts with the physiology of normal sleep, its quantity and quality, the common sleep disorders and their investigation, followed by types of brain injury, its pathogenesis, and functional disturbances before homing in on the link between brain injury and sleep disturbances and vice versa, the importance of good sleep with consequences of lack of same. Then turning to the essentials of good sleep management and non-pharmacological, pharmacological, and other non-invasive treatments of common sleep disorders seen in brain injury. Finally, he discusses the four foundations underlying any specific sleep treatment. There is so much to like about this: in particular the painstaking forensic analysis of the neurochemistry of both sleep and brain injury, the efforts taken to provide very up to date references, the figures and diagrams, and of course the vital consideration of the interactions between sleep and brain injury for their treatment and recovery, and the summary at the end. It is both a substantial academic work and a practical guide.

This truly is *the* reference book on 'Sleep and Brain Injury' and in my opinion should be part of the training curriculum for Higher Specialist Training in Rehabilitation Medicine and in Psychology as well as Neuropsychology, plus a must for clinicians working in sleep services. The abridged version will be invaluable in introducing would-be doctors in medical schools to this subject. It makes for a fascinating read to all practitioners in this field and it will be the bible for 'Sleep and Brain Injury' for years to come.

Dr Rama Shankar Prasad
MBBS, MRCP (UK),
DIPEBPMR,
Consultant in Rehabilitation Medicine

Part I

Sleep

Chapter 1

Introduction to sleep

The basics

This chapter looks at what we know about the basics of sleep. It will introduce the main stages of sleep, highlighting what constitutes each stage. The length, quality, and features of sleep's fundamental importance to humans will be discussed. Different theories of sleep will be explored, alongside reference to the older pioneers such as Kleitman, followed by Dement, Jouvet, and more modern researchers such as Foster.

As I have previously stated this part of the book will, hopefully, already be understood by those whose primary function is sleep assessment and treatment. It is intended to act as a guide for those working with people who have sustained a brain injury.

We all think we know how sleep works, what it looks like, its benefits, and the harms that accrue if we don't sleep. I would argue that none of these statements are wholly true.

Sleep appears to be obvious, the person usually lies down, closes their eyes, hears nothing, they begin to relax their mind and their muscles; occasionally they may move, mutter or even fidget. They appear drugged with the water and thus sleep of Lethe. However, from what we now know, as a result of decades of research, in particular during the 20th and current century, it is far from the mythical dulled and relaxing space we once thought. Moreover, a reduction or even excess in essential parts or stages is health limiting and potentially even fatal.

Sleep is an indispensable part of our everyday existence. It makes up approximately one-third of our lives and is as necessary as food and water. A reduction or excess in sleep affects our ability to learn and store memories. It increases the risk of poor mental health, affecting our ability to have a stable mood making us susceptible to a depressed or anxious one. Furthermore, reduced sleep increases morbidity, including well-known diseases of affluence such as diabetes type II, obesity, hypertension, cardiovascular disease, and fatty liver. Indeed, significant reductions eventually lead to death (e.g. numerous, but see Crowther et al., 2021, for a recent review).

DOI: 10.4324/9780429199066-1

As this book will demonstrate sleep is critical for removing toxins and other accumulated waste from the brain and body. It affects every part of the body from immune responses and disease resistance to specific aspects of memory functioning. In recent years the glymphatic system has become more widely understood, the implications of a faulty one are less well documented. Above all it is dynamic, complex, challenging, and as far from the dullness of Lethe as we can imagine.

The rhythm of life

Humans are driven by a number of basic processes like hunger and thirst. Another drive is the Circadian rhythm or sleep-wake cycle. This rhythm is apparent in every living organism, down to very basic ones such as insects. Some of these creatures have polyphasic, some monophasic, and others biphasic sleep patterns.

Quite why using the word creature compels me to think of Margaret Thatcher is unclear. She strongly and rashly disavowed sleep but later suffered Alzheimer's disease (the link will become very apparent). The truth is no species has evolved to do away with sleep.

Indeed, of all the species studied, the more complex in their evolutionary development, it seems the more complex their need for a particular window of sleep. The parameters within this opportunity for sleep have to operate in a beautifully balanced way for them to be able to work, play, and love well and to the full (probably my only nod to Freud in the entire volume). This applies to humans most of all.

For humans, this rhythm is largely based on the interplay between the brain, body, and the setting and rising of the sun. In that sense, it approximates to 24 hours. I say approximately, as it would be remiss not to mention those pioneers of sleep research Kleitman (see his book *Sleep and Wakefulness*, 1963) and later Richardson, who spent considerable time underground in a cave to prove that we still have our own endogenous circadian rhythm, but that it is increasingly prolonged and disturbed, without sunlight. One of the most profound aspects of this drive is that it is largely (unless cave bound) propelled by parts of the brain and its connections to the endocrine system. Sleep is thus, a primal part of this fundamental cyclic system, as much as wakefulness is. At a given point, it is not possible to have one without the other. It is well known that each of our individual circadian rhythms is different, this can be a small difference or large. Large differences are demonstrated in "morning larks" and "night owls." That is the propensity for some of us to appear to get going when others are lagging and vice versa. In other words, we each experience a different chronotypology. Although largely pre-determined by our genes, our understanding of more recent research into both epigenetics and sleep therapy, as we shall see later, suggests this inheritance is far from immutable.

An introductory chemistry set

The circadian rhythm has a number of prominent features and one separate but important parallel contributor to sleep. One of these prominent features controlled largely by the hypothalamus is core body temperature. A balanced core temperature is crucial for internal vital organs, central and peripheral nerve functioning. As with other functions, it has an optimal range under or above which the body finds it more difficult to operate. Depending on the age and the source of this range, for an adult it is approximately between 35.2 and 37.7 degrees Celsius, in other words the range is quite narrow. What we know is the hypothalamus regulates this temperature to increase and decrease at different time points during the day. The outside of the human body, the skin and body parts further away from the central zones of the brain and thoracic regions, has a shell temperature that is in effect, lizard-like. It is dependent on the external temperature in the environment, primarily. We all have a rise in temperature as the day progresses, indeed it is common to reach temperatures, in our core, of 37.8 Celsius (100 degrees Fahrenheit), in the middle of the day when we are at our most active and alert. Now this does of course depend on whether we are larks or owls; each of us has a slightly different middle of the day. The primary reason for stating this is that it is a common myth that the old-fashioned, for those like me long in the tooth, figure of 100 degrees, is a fever. It is also to underscore the narrowness of the range; perhaps only 0.5 more degrees, depending on the scale, would constitute the beginnings of a true fever. During the day we need a relatively higher temperature for working and functioning well.

It is mainly the reverse at night when the core temperature is instructed to lower. A necessary precursor for sleep itself is for the core temperature to fall to its second-lowest point in the circadian rhythm.

The lowest point in core body temperature occurs in the final phases of sleep, usually in the last few hours prior to waking. Given the narrowness of the range discussed, it is startling to note that the core temperature changes by an average of two (2) whole degrees during the course of the early evening through the morning phase of the circadian cycle. Within the sleep stages, non-rapid eye movement (NREM) sleep is the coolest and rapid eye movement (REM) sleep is the hottest. One hypothesis is that the hypothalamus which, as has been shown, is responsible for temperature regulation, needs to rest and recover as well. It largely switches off during REM. What this means is that external temperatures have a more profound effect during REM than at any other point during sleep. This environmental facet will be considered in more depth during a later part of the book. Indeed, as will be shown it is a fundamental part of early treatment in certain types of brain injury.

Another of the prominent structures, of the circadian rhythm, is the suprachiasmatic nucleus (SCN), which I shall discuss in more depth later.

Suffice as to note here, it is the biological clock. It is located above the crossover point of the optic nerve and notes repeatedly the amount of light entering each eye for later processing in the occipital lobe. Part of the signalling (post notation) occurs through the hormone melatonin. Soon after the sun is beginning to set, the suprachiasmatic nucleus, in response to the light changes, instructs, via the pineal gland, to release melatonin. This hormone builds up during the evening and provides the setting for us to attempt to sleep. Peak plasma concentration is usually around three to five hours after sunset or darkness occurs. Although, interestingly this peak is much later in certain sleep disorders, such as obstructive sleep apnoea. This partially explains why so many prior to continuous positive airways pressure (CPAP) treatment have become more night owl like, even though they may have started life as larks (see Barnaś et al., 2017; for a wider discussion on its relation with metabolic disease see Song et al., 2019). Melatonin does not generate sleep directly but suggests when the window or sleep opportunity should begin. This increasing concentration peaks later in the evening and as sleep continues (if successful) during the night it begins to reduce.

As the morning light changes and enters the eyelids through to the eyes and then on to the SCN again, it commands the pineal gland to halt its release. In addition, the levels of the hormone cortisol usually rise gradually as we sleep. The night progresses to peak cortisol in the morning, and then as the day continues it gradually diminishes, to be re-set once again during the night's sleep. Whereas melatonin shows us when the process of sleep should begin, cortisol levels are high in the morning to help us wake, and then diminish to assist sleep by their relative absence at night. This is all part of what has been termed Process C or the circadian clock. In effect Process C sets thresholds for falling asleep and waking up (Daan et al., 1984.)

Process S (the parallel process of sleep and wake) may be said to be the homeostatic process, or sleep propensity (Beersma, 1998.) In this regard, a different chemical, adenosine, begins to have an effect on the system. Whereas melatonin and cortisol may be delineated as both hormones, adenosine is more clearly a neurotransmitter. From first waking, it builds up throughout a day. Peak concentrations of this chemical circulate some 12 to 16 hours after first waking up. At this point, adenosine has stuck to the maximum number of adenosine receptors in the brain which then compel the person to sleep. Adenosine is in its turn driven by the gradual depletion of the body's glycogen stores. These stores are the powerhouse of our energy. Glycogen utilization is an essential component of the glucose cycle. Without this process growth and other developmental necessities do not occur or at least are problematic, ranging from muscle atrophy through to death (e.g. Peng et al., 2020).

In essence, whilst Process C is almost a relentless clock, analogous to a nuclear one; Process S is mutable, affected by human behaviour (including caffeine intake, Figure 1.1).

process S

process C

additional sleep
pressure

Increased SWA

Increased TST

sleep   work   sleep

7   23   7   23   7   23   10

time of day

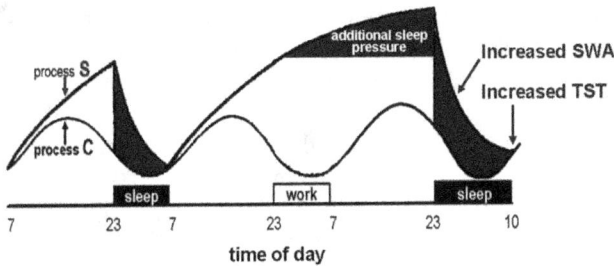

Figure 1.1 Process C and Process S

However, if both are working in harmony a good sleep-wake cycle would likely occur. I say most likely, as new hormones and neurotransmitters affecting both the sleep and wake cycle are being discovered. One such is the neurotransmitter orexin and its sibling hypocretin. These are not, as was first thought, different names for the same thing. The former refers to protein products, the latter to genetic products; both were not discovered till the late 1990s. Both are intimately linked to food intake and wakefulness. The orexin system, in particular, appears to act as an integrator of metabolic and circadian cycles and processes. One of the central features of orexin is its ability to switch on both lateral aspects of the hypothalamus. This subsequently promotes wakefulness in the brain stem and thence the sensory processing of the thalamus. These discoveries have had a profound effect on our understanding of narcolepsy. At post mortem one of the startling findings in narcoleptic brains was the great, almost total loss of cells that would normally be expected to produce orexin (Mieda and Sakurai, 2016).

The sleep world has four or is that five stages?

Kleitman, once again, was involved in later pioneering research through one of his doctoral students, Aserinsky. In 1952 the latter discovered that infants he was observing appeared to have two kinds of sleep patterns that were repeated during each night. The first involved eye movements and the second did not involve any at all. He along with the former named these two phases non-rapid eye movement (NREM) sleep and rapid eye movement (REM) sleep. Later research by Dement, Kleitman, and Aserinsky found that the brain activity when awake was virtually identical to that found during REM. Moreover, they also seemed to have found a clear link between REM sleep and dreaming. During this period when asleep, the body is almost perfectly still, yet the heart rate will escalate, breathing becomes increasingly shallow, our minds are active (in dreaming and otherwise engaged, e.g. memory work) and of course the eyes move constantly and rapidly.

The other phase (NREM) is subdivided again into at least three and in older understanding four stages. Stages one and two may be described as moderately light sleep and stages three and four are the deepest period of sleep. Stage one is felt through drowsiness and a relaxing of muscle tone. Stage two is considered light sleep; muscle tone is significantly reduced together with a slowing of the heart rate. Stages three and four could be seen as degrees of deep sleep. It is here that the heart slows even further, blood pressure drops, slower brain activity is recorded, muscles are by now fully relaxed, and there is a further decrease in body temperature. Crudely, these stages and their depth are defined by the lack of REM together with the supposed difficulty in rousing a person from them. Thus, stage one has no REM but is potentially the easiest to rouse from (in NREM) but stage four, whilst also without REM is the most difficult to rouse from. These, by now four, stages together with the cycles through the night are known as the architecture of sleep. Today NREM is denoted as N1, N2, and N3, the latter being a subsumed version of the old 3 and 4, and now considered simply deep sleep. It is important to understand that humans have the longest cycle length of any mammal; approximately 90–110 minutes, and that we repeat through the four stages on average five (5) times each night, barring disruptions. It is widely misunderstood that each stage is almost a replication of the last, it isn't. To begin with the first cycle, although traversing from NREM to REM, dwells in the depths, the second is similarly found submerged in primarily stages three (3) and four (4). Then the transition of cycle three happens in which far more REM sleep is found. Indeed, near to waking the most REM sleep is frequently in evidence. The problem here is, undoubtedly, the descriptors; a muddling can occur between rhythm, cycle, and stage. Suffice to say four stages (one REM and three NREM combined) are repeated to different degrees within five cycles, held captive by the diurnal variation of the circadian rhythm (see the standard sleep graph, known as a hypnogram in Figure 1.2).

The other part that needs noting is that all NREM and REM are indispensable for physical and mental well-being. Prolonged deprivation of one or the other will lead to increased morbidity but also ultimately death (numerous, but see Tilley et al., 1992 for a succinct review).

When we say all species, we mean as far as we can demonstrate through behavioural observations of smaller species (bees, beetles, even, molluscs, and so forth) and actual polysomnography results from larger species (e.g. great apes.)

Although common across species, differences emerge between the amounts each needs. It would appear, and it is my contention that there is greater mileage in the theory, that the more complex the nervous system, the more the quantity of sleep is needed. That is to say generally, the degree of concordance breaks down with unusual phyla. Bats have been cited frequently as one outlier, as supposedly needing an inordinately long amount of sleep relative to their size. However, if you glance at some of the literature

Hypnogram

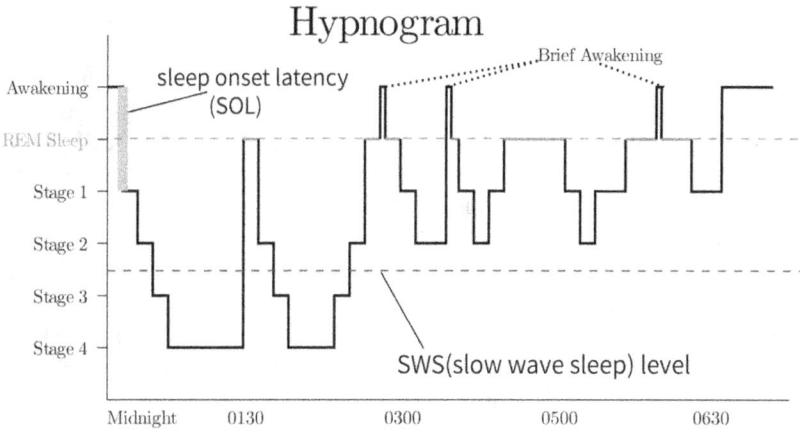

Figure 1.2 Sleep hypnogram

from the studies at UCL (University College London e.g. Smaers, 2012) on body size and brain size, more obvious relationships to body size frequently occur than simply brain size. It appears that bat manoeuvrability was paramount in their evolution (think of their darting, rapidly changing flight to catch insects) than larger body size *per se*. This is in contradistinction to most other carnivores (humans included) who have increased body size almost perfectly correlated with their increasing brain size.

The more important point here is the universality of the need for sleep and a general trend. This brings us to the requirements of humans who have the most developed of all nervous systems, of all species on earth. Other species may have larger overall brain size with important differences. By that not simply brain size, but, most especially, frontal lobe size and overall complexity of the attached nervous system are key to understanding this. We have the largest frontal lobes of any species and they need a lot of maintenance.

Another distinction is REM sleep; it is only more evolutionarily developed species that experience this. Unsurprisingly, it is humans who display the largest propensity for this particular form of sleep, indeed when compared even to our closest cousins, the great apes and chimpanzees; we experience nearly three times as much each night. It is also the quality of this REM that is different; it is somehow more densely packed. Walker (2017) presents an elegant notion that it is the link with our becoming an erect and non-tree dwelling species that (in addition to other interesting ideas) propelled this development. In other words, REM came out of our evolutionary development. I would argue that it is rather our frontal lobe development that was the precursor, to this enhanced REM sleep. This could, of course, work hand in hand with becoming erect as a species, but it cannot be the sole reason.

This argument is a two-fold one. On the one hand, our giant (in comparison to other species) frontal lobes, need nightly servicing, deep sleep satisfies some of this demand but it is only REM that completes it. On the other hand, the increased quality and quantity of this REM sleep enhanced frontal lobe development. Evidence of sorts can be found in the work of Boysen (2006 amongst others). She found during a process called enculturation, that those chimpanzees which had a larger amount of human interaction and training (unwittingly and wittingly to be more human,) demonstrated changes in the pre-frontal cortex, more akin to human development than chimps. Chimp brains, although having a pre-frontal cortex, are driven more by the primitive limbic structures and lack to a significant degree the modifying pre-frontal systems that can actually harness drives, which humans are able to do. As such chimps are slightly less cooperative and social than humans. The inhibitory, marshalling, managing, executing, and skilled socially cooperative human behaviour is largely (although not solely as we shall see later) found in our giant (*sic*) frontal cortex. It is REM sleep that is needed to feed and I would argue "stabilise" this lobe. Disturbances, as we shall see later, in REM and/or frontal function have grave consequences for our mental and emotional well-being.

The quantity and quality of sleep can be strained

First, each of us needs the opportunity to sleep enough. Too often patients, when discussing their "insomnia" with me, fall at this first hurdle. It is also, in my experience, the hurdle that most clinicians fail to even ask about, concentrating as they do on all other aspects to its (opportunity time) detriment. Sleep opportunity time is deceptively simple. Have you budgeted enough time during the course of a 24-hour day to include enough time to at least *attempt* sleep? The answer in many cases is no. Thus, if you repeatedly go to bed at 2am in the morning and then try and get up at a "normal" time in the morning, say 8am, your total sleep opportunity time was 6 hours. Now, if you go to sleep within seconds (unusual), progress perfectly through the stages and cycles (very unusual), as previously described, and then wake at 8 am on the button, you will have enough sleep to function but not enough to feel fully refreshed. There were also an awful lot of ifs in that last sentence. More likely, is some form of a delayed onset when going to bed, an average amount of small wakeful periods during the night (between 30 and 80 minutes in total) and an impromptu waking by your partner, refuse collector, lark, and so forth. Meaning, in effect, you actually ended up experiencing only about 4 to 5 hours of real quality sleep, which is definitely not adequate in the short or long term. As we now know slightly inadequate amounts can build into significant sleep deficits over a relatively short period of time. In conclusion, we each need a basic opportunity or window of time in which we could possibly experience sleep.

If we examine briefly the amount in each stage that is required as opposed to the total overall, again we discover healthy (and thus, non-healthy) ranges. Although as with other parameters discussed, these figures depend on age. Thus, very young children appear to need and enjoy far greater time in REM sleep as opposed to teenagers whose deep NREM sleep is of paramount importance. Between 55 and 97 Minutes for 7 and between 70 and 124 minutes for 9 hours of sleep is considered the optimal amount(s) for deep sleep. So, what happens if you sleep less or more than these requirements; let us say broadly the range is between 55 and 125 minutes each night. Deep sleep is fundamental for repair of the mind and body. It is during this period that the body slows everything down; heart rate, blood pressure, temperature, and movement are all reduced. It is essential for memory consolidation and new learning. Deep sleep deprivation propels the deprived into a cognitively impaired state. Indeed, recent studies have found that the protein *tau* associated with Alzheimer's disease is raised in those who have higher rates of sleep disturbance. Even more powerfully in this regard is that it was most elevated in those patients who have the poorest NREM deep sleep. Using spinal fluid assays which measured amyloid β kinetics, they concluded that disrupted deep sleep increases Alzheimer's disease risk; not the other way around (Holth et al., 2019).

The opposite is also true, in that, better deep sleep is associated with better health. As we shall see later, part of this is to do with the restorative processes that are accomplished during this period of sleep. It has been argued that although some aspects of brain function are dependent upon deep sleep, it is much more useful to think about it as essential for our overall physical health; including the brain. It cleanses the blood, the body replaces dead cells, and it heals wounds and builds muscle. There is no single physiological function or organ that is not affected by deep sleep or the lack of it (see for example: Calhoun and Harding, 2010; Mullington et al., 2010; Meltzer et al., 2020).

Furthermore, blood pressure fluctuates between 30 and 40 mmHg during the 24-hour day. The American College of Cardiology's 68th Annual Scientific Session presented an important paper regarding those who napped in the day around lunch. They reduced their blood pressure by around 5 mmHg (roughly equivalent to a standard anti-hypertensive medication). Moreover, for every 60 minutes of lunchtime sleeping average blood pressure dropped by 3 mmHg for the entire 24 hours. Accompanying such changes are those related to heart rate. The range of decrease in heart rate whilst sleeping is, once again, age and sleep-stage-dependent. However, when taken as a whole heart rate decreases between 14 and 26 beats per minute. During sleep it can range from 40 to 100. In deep sleep, the average is 55 and during REM approximately 75. Overall, resting heart rate (including that during sleep) is a good predictor of cardiovascular health. In their most recent review Bonnar et al., (2018) argue that slightly prolonged sleep and better sleep management (hygiene) strategies enhance cell repair and improve recovery times. As powerfully, they argue that these also improve performance even in elite athletes.

Conversely, REM sleep is a highly active phase of sleep; brain activity resembles being awake, heart rate can become extreme and dreams are vivid and prolonged. Figures of between 20% and 25% of the total time asleep have been found to be optimal for this stage of sleep. Between 84 and 105 minutes for 7 and between 108 and 135 minutes for 9 hours sleep is considered the optimal amount(s) for REM sleep. So, what happens if you have less or more than these REM requirements; let us say broadly the range is between 85 and 135 minutes each night. There is an equal debilitating effect from REM sleep deprivation and excess. This is unlike deep sleep, which appears to have no upper limit. REM sleep is more associated with mental health and well-being. It is the healer of emotional wounds. Germain (e.g. 2013) in a review demonstrated that although a specific drug which reduces noradrenaline did not affect sleep quality or improve PTSD symptoms, REM sleep is affected by the symptoms themselves. Of even greater fundamental impact on our understanding is the work of Wassing and his colleagues (e.g. 2018 and 2019.) REM sleep changes neuronal representations of traumatic, stressful, and anxiety-provoking experiences. It seems that more recent emotional disturbances activate the limbic system in those with REM sleep disorders, more than older memories of emotional distress. The REM sleep duration appears as a discreet predictor of our ability to process emotional problems of the day. Those who have REM sleep deprivation are less able to process such information and ' ... downregulate emotional distress' (Wassing et al., 2019).

As Laura Palagnini has argued, where once depression was seen as causing increased sleep generally, extensive polysomnography studies appear to show at a certain point it is a prolonged REM specifically, which is causative of depressive relapse. In one paper she and her colleagues argued that prolonged REM, above a certain amount, may well be a true endophenotype of depression (Palagnini et al., 2013.)

It is also well known that REM sleep disturbances are frequently found in Parkinson's disease. These disturbances can be seen as independent predictors of disease progression and produce a number of changes to memory, executive function, attention, emotional processing, and sensory perception. Gjersted and her colleagues have suggested that specific REM sleep disorders may be predictive of early cognitive decline and argues forcefully that 'REM sleep behaviour disorder (RBD) is recognized as a hallmark for the development of α-synucleinopathies.' Their later contention is that potentially specific REM sleep disorders may be predictive of Parkinson's disease more precisely, although recognizing that these have not been developed (Gjersted et al., 2018.)

As I shall show later this research has implications for anxiety and mood disorders, as well as frontal lobe function. In addition, it points to the wider link with REM sleep and its particular benefits and problems.

You can have both too little and too much of a good thing

Having given oneself the benefit of an adequate sleep opportunity, the next question may be what the "correct" amount within that opportunity time is for actual sleep to occur. This is complicated by the knowledge of the different needs at different ages across the human lifespan. Thus, it is largely true that the younger the age the greater the amount of sleep needed. On average a newborn will need between 14 and 17 hours, infants (12–15 hours,) toddlers (11 and 14 hours,) preschool-age children will need between 10 and 13 hours. Requirements change again for school-age children, but there is still a need for between 8 and 11 hours each night, depending on the age of the child, then teenager. As we shall see 7 to 9 hours for young adults and adults, and 7 to 8 hours of sleep for older adults are considered the healthiest ranges of actual sleep duration.

Thus, it is now a largely uncontested idea that the normal variance for healthy sleep, is somewhere between 7 and 9 hours each night for an adult. Indeed, go too far outside these parameters and similar health problems seem to accrue, at either extreme.

Evidence began emerging strongly in the early 1990s (e.g. Tsubono et al., 1993) that sleep duration was one of the key variables in the risk of developing a wide range of so-called diseases of affluence. Later meta-analyses and reviews from around 20 years later (e.g. Cappuccio et al., 2010) began to elucidate not only the greater risk of developing such diseases but also a more specific understanding of the exact duration needed, either above or below the average ranges to develop such disease. Indeed, Cappuccio et al., (2010) were able to show that the relative risk (RR) of developing type II diabetes, if you slept for a short time (defined as less than 5–6 hours each night on average) was 1.28, for a longer duration (defined as greater than 8–9 hours each night on average) was 1.48. As interesting, or worrying, was the finding that particular qualitative indices were even more important at increasing the RR. Two examples will suffice, difficulty initiating sleep was associated with an RR of 1.57 and difficulty maintaining sleep was undoubtedly the worst, its RR was found to be 1.84. In a summary of their conclusions, those who had difficulty maintaining sleep had almost twice the probability of developing diabetes type II, than those in the lowest group. There are many other implications for too little and too much sleep, and it is not the purpose of this book to dwell on these; only as we shall see for those with a brain injury. The interested reader should look to authors such as Cappuccio and later summaries by The National Sleep Foundation of America (Hirshkowitz et al., 2015).

A recent study by Lubetkin and Jia (2018) found that sleep deprivation or excess, again, has slightly different effects on mortality, as opposed to the above morbidity rates. This study was interesting for a number of reasons, not least of which was the large cohort it studied (n=2,380) together with the extended follow up. Initial data were gathered between 2005 and 2008, then the cohort was re-examined in 2011. Of those studied over the age of 65 years, a clear differential was made. The shorter the duration of sleep the greater the morbidity. However, the longer the sleep excess the greater the adverse mortality. The mean quality of life scores for those participants sleeping more than 10 hours were the most significantly lowered and produced the highest mortality sub-set. It begs the response perhaps that shorter duration during earlier years was associated with greater morbidity, which translated in later years to excess sleep needs (potentially related to the accumulated morbidity) and thus potentially higher mortality rates. In any event, what is apparent is that the problem of being too far outside of normative range had (and has) deleterious effects on the individual.

Thus, for different reasons too much and too little sleep is both dangerous. Specific lengths and types of qualitative problems can also promote certain diseases over others. As I shall argue later, this fits with another parameter of the human body and its needs; we have, at least for some functions, an optimal zone, period, or range, above and below which functions can go awry.

Chapter 2

Introduction to the neurochemistry and neuroanatomy of sleep

As I have touched upon thus far, sleep is a carefully orchestrated affair, which involves a great many inter-related neuroanatomical and neuro-chemical substrates.

In essence, as Lugaresi, Provini, and Montagna (2004) argue sleep itself is a co-ordinated caudal to rostral process, which is then reversed partly by the frontal lobes to become a rostral-caudal process, but primarily the brain-stem and mesencephalic structures are important for waking; enacted en-dogenously and through light sampling via the suprachiasmatic nucleus. As stated earlier different sets of neurotransmitters and hormones affect each process to produce either wakefulness or sleep.

Although many neurotransmitters have been discovered as important in sleep, more are being found and subtleties in understanding are continuing to evolve. What is presented is the author's view of the current knowledge re-garding these. It should be noted that this is not intended to be a book about neurotransmitters, there are over 40 different types of neurotransmitters and each as it is discovered seems to point more to its interplay with other neu-rotransmitters than being an "answer" to something profound and in isolation. Our understanding of these individually is getting better; our understanding of their interplay in interacting systems is still in its infancy.

During wakefulness, neurons sensitive to histamine, dopamine, serotonin, and noradrenaline are activated. These in turn, inhibit the sleep-promoting neurons of the ventrolateral preoptic nucleus (normally VLPO occasionally VLPN). In contradistinction, γ-aminobutyric acid (GABA) cells in the basal forebrain and anterior hypothalamus play a major role in inhibiting these previously noted neurotransmitters and hormones. They flood the brain, inducing sleep by inhibiting cells involved in arousal functions. This in-hibition particularly affects the cholinergic system and its inhibition through primarily GABA, this then largely switches off the cortex; especially that of the forebrain. GABA cells, potentially uniquely, are highly active during NREM sleep and less so during REM sleep. Siegel (2004) attests to the crucial role of the anterior hypothalamus in sleep (and to some extent sleep and wake), whilst the posterior hypothalamus is concerned more with

DOI: 10.4324/9780429199066-2

wakefulness. In large measure, deeper centres could be viewed as being more responsible for wakefulness; these are in turn inhibited, modified, and dampened through higher areas, which include the anterior hypothalamus, forebrain, and frontal lobes. GABA-ergic neurones affect the cholinergic system and the histamine cells. Much of the widely available medications used to treat insomnia enhance the GABA system. For example, Zolpidem enhances GABA action on GABAα receptors. In related fashion Zolpidem also affects histaminergic neurons which then produce analogous NREM sleep, which resembles natural NREM sleep.

Whilst this description appears straightforward, our primary basis for understanding the hormone and neurotransmitters in sleep and wake is through the work undertaken in understanding adenosine. It is the breakdown product of cell metabolism, by the end of the day large amounts of cellular breakdown has occurred and thus an accumulation of adenosine is apparent. It is predominantly released during wakefulness; indeed, the greater the length of time a person is awake the more is produced. Adenosine is a complex paracrine neuromodulator. In other words, it acts in a local manner unlike endocrines, which affect whole systems. It inhibits the effects of acetylcholine and glutamate, which are largely responsible for keeping us awake

Acetylcholine is critical for vigilance and commencement of cortical function. The pontine mesencephalic reticular formation and key connections to the thalamus, in the forebrain and lateral hypothalamus, in addition to other structures in the basal forebrain; particularly the nucleus basalis are the cholinergic neurons affected by acetylcholine. In contrast to the cortical activation, in the day, acetylcholinesterase inhibitors induce REM sleep. This is also the time when other excitatory neurotransmitters are inactive.

As stated previously GABA is the pre-eminent inhibitory neurotransmitter of the central nervous system (CNS). During slow-wave sleep, GABA is released at a higher rate in the posterior hypothalamus, than during wakefulness and REM sleep. Primary wakefulness structures are based in the posterior hypothalamus, brainstem, and forebrain. The brainstem reticular formation and locus coeruleus have GABA-ergic neurons and these inhibit wake-promoting neurons, and are as a consequence, active during sleep. Within the thalamus, and this structure's connections with the cortex (and thus cortical activation) the inhibitory function of GABA is essential for both starting and maintaining slow-wave sleep. This is of course why Baclofen (as a GABA receptor agonist) amongst other commonly used drugs in spasticity management can not only enhance slow-wave sleep but also induce sleepiness and memory problems at higher doses in patients.

Noradrenaline is unusual, as it can have either inhibitory or excitatory effects. This depends on the kind of receptor involved. Some, such as the antagonist Prazosin, blocks post-synaptic receptors and helps induce sleep onset. Whilst other antagonists, such as Yohimbine, delay sleep latency.

Two contrasting drug effects should be noted. Clonidine appears to inhibit noradrenaline release and thus, produces hypnotic effects. In contrast, stimulants such as Methylphenidate block the uptake and thus increase the concentration of noradrenaline, which enhances wakefulness. Indeed, it is used (underused as I will show later) in the treatment of hypersomnia and on occasion for fatigue problems classically associated with acquired brain injury.

Another neurotransmitter largely responsible for wakefulness is glutamate. Other than high concentrations found in the reticular formation, its other primary base is the thalamus and its projections to the higher cortical areas. As Sherman (2014) has argued understanding is often oversimplified regarding this neurotransmitter but: 'Their evident importance in functioning of thalamus and cortex makes it critical to develop a better understanding of how these receptors are normally activated, especially because they also seem implicated in a wide range of neurological and cognitive pathologies.'

This better understanding is best exemplified by a more recent paper by Bathel et al. (2018) in which she examines the age-old idea of too much of either inhibitory or excitatory neurotransmitters in the pathology of neurological disease. What makes this paper special is its forensic examination of glutamate specifically (and the relationship to GABA) in regard to local hyperexcitability in the occipital cortex and right thalamus of those who suffer from migraines versus healthy controls. Although no significant group differences were found, it is the paper's examination of the current ignorance of these fundamental neurotransmitters that is of interest. Given the complacent view that we know the basics of operation, the conclusion would be that after decades of supposed knowing, we still have a huge amount to really know.

Another equally unknown well known inhibitory neurotransmitter is dopamine. Potentially confusing, this neurotransmitter can also be excitatory. We do know that dopamine-containing neurones can be found in the substantia-nigra and the ventral tegmental area that innervate the hypothalamus, the nucleus accumbens, and significant areas of the frontal lobes. We also know that glutamate and these brain areas play important roles in wakefulness. Moreover, when aroused during wakefulness and REM sleep dopamine neurons are firing actively. Methylphenidate and cocaine alter dopamine uptake which leads to increased wakefulness and increased cortical activation. These were commonly used to treat narcolepsy as a consequence. Conversely both older (typical) and modern (atypical) antipsychotic medication are largely (although they also affect other neurotransmitters) dopamine antagonists. In effect, these drugs block the same dopamine receptors previously identified. In the 1960s the so-called dopamine hypothesis became the largely uncontested neurobiological view of the primary causation of schizophrenia. However, repeated decades of research and criticism have meant that we now understand that although for certain sub-types of psychosis this hypothesis is useful it is only a very partial explanation. Howes et al. (2015) explained, this has been enhanced by a better understanding of the dual role that dopamine can play, for if an antipsychotic were to only work in one direction (most, if not all) then the effects on a dual-acting neurotransmitter (both excitatory and inhibitory) would be impoverished, as the hypothesis itself is only unidirectional. As importantly, their findings do provide a more nuanced support for a specific dopamine hypothesis, for the purposes of this chapter, they also suggest glutaminergic dysfunction is one of the primary likely precursors to the dopamine problem. Indeed, the more we understand the complexity of neurotransmitter systems, the better we are able to explain these phenomena and potentially, the better able we may be to help people with these problems. For a good overview of this more complex understanding see, for example, the chapter by Gründer and Cumming in *The Neurobiology of Schizophrenia* (2016).

Linked to the latter hypothesis and other forms of mental health problems is the inhibitory (but sometimes not) neurotransmitter serotonin. Serotonin neurones in the dorse raphe specifically in the dorsal and median raphe nuclei primarily send their projections to forebrain regions. Thus, once again we see the frontal systems interplay with deeper brain structures being fundamental to wakefulness (and their equivalent sleep). Serotonin itself is linked to wakefulness. Serotonin firing neurons conversely decrease during slow-wave sleep (SWS) or deep sleep but interestingly, unlike some 'wakeful' neurotransmitters, cease completely during REM. This latter knowledge alone is of immediate interest to any clinician who has worked with those suffering depression who nearly always have co-morbid sleep problems. REM sleep, in particular, is seen as essential for 'good mental health' and it is more common for those suffering depression to be affected by early morning waking. This is more likely to reduce REM sleep than any other stage. However, as discussed earlier too much REM is also associated with mood problems. This potential contradiction is an important one, as will be examined fully later. This is also of interest as those modern drugs targeting depression largely affect serotonin systems. These same drugs are often associated with REM sleep suppression, yet also can lift mood. Frequently, insomnia treatments such as the Z drugs (e.g. Zolpidem and Zopiclone) and the benzodiazapines, have to be prescribed alongside these drugs for a significant portion of the depressed population (Wichniak et al., 2017).

Once again when viewing the primary regions of the brain that are key for sleep and wakefulness the hypothalamus is involved and a very important neuromodulator (on occasion known as a neurotransmitter, but technically not necessarily!) that of histamine. Like other neuromodulators these wash around in the cerebrospinal fluid for very much longer than standard neurotransmitters and in turn affect these and others in a bigger way, they could be viewed as super system modifiers. However, and rather confusingly, they can also be neurotransmitters which act directly and specifically in the short term. Returning to site-specific regions, the tuberomammillary nucleus in the hypothalamus is the only part of the brain that produces histamine. This is produced in high quantities during the day and less at night. In addition, extracellular histamine levels in the frontal cortex are positively correlated with the amount of wakefulness (Chu et al., 2004). Higher levels of histamine have also been shown to be important in inflammatory responses, amongst other key functions. It is thus essential for alertness and wakefulness. During sleep GABA in the VLPO inhibit histamine neurons, in related ways anti-histamines, most especially first-generation ones, in addition to reducing hay fever and other allergic reactions produce profound sedative effects due to the dampening of the histamine system.

Although previously mentioned, orexin (hypocretin) is found in the locus coeruleus. In this chapter, it is of more interest to note that it is also found in large quantities in the tuberomammillary nucleus, which then has dense axonal projections to the cerebral cortex. As well as sleep, this area has been invoked as critical for learning and memory, which will be discussed later. In addition, and unsurprisingly, given the foregoing, both the tuberomammillary nucleus and the locus coeruleus are essential in maintaining arousal.

Many other neurotransmitters, modulators, and hormones could potentially be discussed (including, galanin, growth hormone, and thyrotrophin) but, only two more will be considered, to illustrate the area under consideration, one for its novelty; the other for its ubiquity. Indeed, as the book progresses the latter will become a central one, these are vasoactive intestinal peptide and melatonin.

Vasoactive intestinal peptide (VIP) was discovered almost half a century ago (Said and Mutt, 1970), it is less well known than most of those previously discussed and yet has profound links with GABA systems and the all-important master circadian pacemaker of the brain; the SCN. The connection with GABA is interesting. In their interventional study, Korkmaz et al. (2010) systemically administered VIP in Parkinsonian rats. They examined a number of potential mechanisms of action; the two primary ones were activation in the ventral anterior nucleus of the thalamus (VATh) and growth factor in mast cells in the brain. They concluded that the protective effect of VIP on motor performance is most likely related to GABA increases.

This runs parallel with other research which demonstrates the neuroprotective and anti-inflammatory effects of VIP. It has even more fascinating effects on sleep. Seminal work undertaken particularly in Mexico (e.g. Prospero-Garcia et al., 1993; Jiminez-Anguiano et al., 1996; Prospero-Garcia et al., 2011) demonstrated repeated evidence in cats, rats, and mice that VIP has highly specific functions in regards to REM sleep. It would appear that VIP accumulates in the cerebrospinal fluid (CSF) as a consequence of being awake, indeed the more prolonged the wakefulness the higher the concentrations of VIP. This is akin to the wakeful accumulation of adenosine levels. Through a series of experiments involving extraction over systematic time periods, CSF and its range of VIP matched precisely the amount of REM sleep deprivation. Not only this but rebound REM sleep is reduced following the extraction (and thus the VIP). Finally, later work (Simon-Arceo et al., 2003) demonstrated long-lasting improvements in REM sleep after microinjections of VIP into the amygdaloid central or basal nuclei; with the former showing the more robust effects. As with other neurotransmitters, modulators, indeed, wider aspects of physiological function, caution needs to be taken here, remembering one of the themes of the book. Too much production of VIP as seen in the rare endocrine tumour; VIPoma, produces devastating effects on a number of interrelated systems, including but not limited to: chronic diarrhoea, hypokalaemia, hypercalcaemia, and hyperglycaemia.

Finally, in this brief overview of some of the more important sleep hormones, neurotransmitters, and modulators, probably the one most well-known, and potentially most important in current understanding: melatonin will be further examined.

Melatonin is, as stated in an earlier chapter, the provider of the window to sleep, its waxing and waning is essential for a healthy circadian rhythm and regulates the sleep-wake cycle. It is secreted by the pineal gland at night after the signals received from the suprachiasmatic nuclei have interpreted the available light from the optic nerves. Its relevance to the circadian rhythm is thus well established. It has therefore become the medication of choice for a wide range of sleep disorders from simple jet lag through intractable insomnia. Particularly as it has, in comparison to the benzodiazepines and other anti-insomnia drugs, fewer side effects.

What is less well known is its wide-ranging other effects. Melatonin is an endogenous hormone, which has been described as a multi-tasking molecule with a great potential for other applications. It is involved in mood regulation, sexual behaviour, immune modulation, and anti-inflammatory responses. It alters the effects of both pro and anti-inflammatory cytokines, in different conditions. Maestroni and colleagues (2001) argued that it has such potential for fighting disease; it may in time be considered for viral, bacterial, and even cancer treatment. In this regard, it is interesting to note its relative decline with age. Age being a time of relative sleep impoverishment and increasing morbidity. Indeed, Tarocco et al. (2019) argue in their review that it may be considered a: 'master regulator of cell death and inflammation' and it is this relationship, which is one of the most important ones, that will be made in regards to brain injury and its recovery later.

Neuroanatomy basics

Thus, in all respects sleep itself cannot be split from wakefulness. They are two sides of the same coin. Certain key neuroanatomic regions have previously been discussed, now highlights of other areas helpful either to sleeping or wakefulness will be outlined. Indeed, much of the earlier work on Processes C and S has been criticized for being too phenomenological and not based on rigorous neurophysiology. Step forward the reciprocal inter-action model and the flip-flop switch models.

The reason for highlighting these models is that they demonstrate a number of key principles. First, there is no complete sense of sleep, wake, or the neuroanatomy of either of these. Second and relatedly, a number of complex models have been proposed; these involve systems and remain largely more or less robust hypotheses, not definitive answers. Finally, it shows that although sleep has been with us as long as we have been, that it cuts across all species; it is only relatively recently that it has been system-atically studied. To give a better idea of this, the first textbook in the USA was acknowledged to have been published in the 1980s. The first properly organized postgraduate training is still being developed in the USA. In the UK, it is even worse, in psychology, neuropsychology, and medicine, the most fundamental sciences associated with sleep and its treatment: sleep training in undergraduate and post-graduate programmes are either non-existent or a matter of a few hours, at best, in total.

Returning to the main theme, it is commonly held that, since the late 1940s (Moruzzi and Magoun, 1949) there has been an understanding that the as-cending reticular activating system (RAS) which sends projections from the brainstem to the cortex, is the awakening and arousal portion of the brain. The classic paper describes the essential caudal to rostral generation of wa-kefulness. It is also true to say that as far as we are aware the majority of the most important neuroanatomical substrates responsible for sleep and wake-fulness are found in the brainstem, hypothalamus, thalamus, and their pro-jections and connections to the frontal lobes. The coordination of either ascending or descending pathways is light sampling undertaken by the SCN relayed to and then processed by the thalamus. Later work (for an overview see for example, Boly et al., 2012). demonstrated how it was the frontal lobes and their connections with the lower order, (phylogenetically and neuroana-tomically) deeper regions in the brain which were incredibly active during wakefulness together with similar patterns during REM sleep. Even more critically, what I have tried to demonstrate is that the same neurotransmitter may have distinctly different effects depending on the particular site, and its relation with that site, in the brain. In the posterior hypothalamus, GABA promotes sleep at certain times (Lin et al., 1989). On the other hand, the same neurotransmitter activating the pontine reticular formation will engender wakefulness (Xi et al., 1999). It is also more usefully acknowledged, although crudely, that generally GABA is an inhibitory neurotransmitter. The reason this is useful is its dampening effect on the cascade of neurotransmitters that occur during wakefulness, and how this helps propel sleep.

In 1937 the French neuroscientist Bremer hypothesized the brainstem was key to understanding wakefulness. Sleep was proposed to be a passive process after he demonstrated that when animals were dissected at the brainstem in the intercollicular region no waking occurred. Indeed, continuous sleep occurred. Sleep was a withdrawal of sensory input. Maruzzi and Magoun (1949) dissected a lower point (C1) which showed that some wakefulness occurred in these animals, but not complete wakefulness or sleep. It was this finding that made them propose an endogenous arousal system (between Bremer's and their own "cuts".) In other words, the ascending reticular activating system (ARAS). Shortly after this Aserinsky and Kleitman (1953) outlined the phenomena of rapid eye movement (REM) sleep in new-borns and older babies. Or as they phrased it: 'regularly occurring periods of eye motility and concomitant phenomena during sleep,' while the cortex is activated yet the person is asleep. Dement (1958) then suggested dreaming was mainly (but not solely) associated with this period during sleep. He went on to draw parallels with waking and REM sleep electroencephalograms (EEGs). Work undertaken at the same time in animals by the great French neuroscientist Michel Jouvet delineated the principles and neuroanatomical substrates of REM sleep. After spending a year working in America with Magoun, he concentrated his efforts on wakefulness, dreaming, and REM. It was he who coined the phrase: paradoxical sleep, for REM. It was he who found that the rostral pontine section of the mid-brain or pontine tegmentum and the locus coeruleus were critical areas for wakefulness and REM neural activity. His team was also helpful in pointing the way to the essential differences between wakefulness and REM and he was the first to propose the kinds of neurochemistry that underlay all of these proposals (1999). Bizarrely largely ignored in conventional textbooks on sleep, he was unquestionably one of the great pioneers in sleep research. A major finding here was that cortical activity in the frontal and posterior lobes is less coherent during REM than when awake. In other words, there is a high degree of regionalized activity, which is specialized to the purpose (REM, awake, deep sleep, and so forth). This has become part of the acknowledged explanation for dreaming (see for example: Antrobus and Bertini, 1992). It was Jouvet who also first cemented the idea of particular neurochemical substrates involved in sleep and wakefulness. Especially that the cholinergic and monoamine neurotransmitters play a large role in the ascending cascade of wakefulness.

The forebrain can be viewed largely from the perspective of NREM sleep and wakefulness, as opposed to the previous discussion on REM and wakefulness. In addition to the thalamic and hypothalamic areas noted, it is the frontal lobes that are important to understanding these states. The work of von Economo, is crucial here, especially his post-mortem observations during the 1920s. From these, he hypothesized various effects on sleep depending on the primary lesion found. Most clearly anterior hypothalamic lesions produced severe insomnia in these patients, whilst the reverse was true of posterior hypothalamic lesions: incredible somnolence (throughout most of the day) was observed and reported (von Economo, 1930). Crudely, this is almost the difference between insomnia and hypersomnia. Thus, the first two essential sleep and wake neuroanatomical hypothalamic substrates were identified, these ideas were supported through many subsequent animal studies (see for example: Nauta, 1946; Swett and Hobson, 1968; and Sterman and Clemente, 1962). As was discussed earlier these areas also map onto GABA-related activity and rich connections with the frontal lobes.

A brief visual guide to the aforementioned discussion is outlined in Table 2.1.

We now know that key neuronal groups in the locus coeruleus (expressing noradrenaline), dorsal raphe (expressing serotonin), tuberomammillary nucleus (expressing histamine), and the pons; specifically, the lateral dorsal tegmentum and the pontine peduncular tegmentum (expressing acetylcholine) are critical for the ascending arousal pathway. This then projects from the deeper brain areas described through to higher cortical areas. We now know that there are a number of key neuronal groups in the VLPO (expressing GABA and galanin) which have rich synaptic projections to the above wakeful neurone groups previously described, which are critical for sleep. These latter projections and neurotransmitters, in contradistinction to the excitatory mechanisms described above, are inhibitory and promote sleep. The idea is that there is a mutually inhibitory mechanism, akin to a flip-flop switch in engineering. This means according to the hypothesis, you are either awake or asleep, as populations of either inhibitory or excitatory neurotransmitter system dampen the other. As we know from the simple conundrum of drowsiness this model is not strictly accurate, however, it is the best fit model we have currently. The idea as to the neuromodulators involved in moving from one state to the other is the three mechanisms previously described. In other words, orexin is seen as the one of the main neuromodulators for instigating wakefulness. The well-known increase in adenosine during the day and its full weight of effect by night. Finally, what we have previously discussed in regards to the basic mechanism of light sampling via the SCN; in other words the circadian clock or rhythm. Acting in concert with more recently discovered neurotransmitters, like VIP, we are gradually

Table 2.1 Some of the important Wake–Sleep promoting neurotransmitters

Wake-Promoting Neurotransmitters	NREM Sleep	REM sleep
Glutamate	GABA	Acetylcholine
Acetylcholine	Galanin	Glutamate
Dopamine	Adenosine	GABA
Noradrenaline	Melatonin	Glycine
Serotonin	Serotonin	
Histamine		
Orexin		

gaining a more complete picture of the complexity these interacting systems have on human sleep and wakefulness.

One final specific neuroanatomical region will now be examined; that of the higher cortical structures, especially the frontal and parietal lobes. Although known for over a century (see for example: Davis et al., 1937), the precise nature and function of these structures, especially when considering something called a sleep spindle has been debated for a long time. Even today only perhaps 30% of the function is entirely understood (Piantoni et al., 2017). Sleep spindles usually occur on the descending aspect of each wave during sleep. They are widely held as important overtures to the end of stage 2 and deeper NREM sleep. Only lasting between half and a couple of seconds, they are deemed essential for memory consolidation, sleep awareness, and thus survival. In abbreviated form, NREM is hugely important for episodic memory consolidation and REM sleep is hugely important for procedural and emotional memory consolidation. Despite this apparent understanding and enormous interest over many decades, clarity around their fully worked through neuroanatomical regions with their exact functions are yet to be completely understood (Caporro et al., 2012). Sleep spindles and their importance will be examined frequently as the book progresses.

Previously, Walker (2017) described the orchestration of deep sleep via the frontal lobes. As if a sound pinged from the front to the back of the brain:

> They are like the sound waves emitted from a speaker, which predominantly travel in one direction, from the speaker outward... And like a speaker broadcasting across a vast expanse, the slow waves that you generate tonight will gradually dissipate in strength as they make their journey to the back of the brain without rebound or return.

Building upon this knowledge of sleep spindles and his beautiful evocation of sound waves, leads us to even more complex, "musical forms." Del Felice et al. (2014) identified the electrical source of sleep spindles and differentiated between slow and fast sleep spindles. They suggested that as well as the frontal lobes, being especially important for the generation of slow spindles, the parietal and limbic structures seemed crucial for fast spindle generation.

Even more recent work by Piantoni and colleagues (2017) built on this and found further evidence of frontal and parietal involvement, but added nothing to the long-standing debate regarding any temporal lobe involvement. This is unlike the work earlier shown by Caporro et al. (2012) which through careful analysis of coterminous EEG and functional magnetic resonance image (fMRI) data revealed temporal lobe sources for the initiation of some increases in spindle density during NREM sleep. They also confirmed the importance of Brodmann area 7 (superior parietal lobe) and found some evidence for frontal (Brodmann areas 9 and 10) in addition to the posterior cingulate area. Key findings across most studies suggest midline frontal, temporal and parietal regions, in addition to the core spindle generating activity bilaterally in the thalamus, are the most frequent generating areas found. Connections between these cortical and thalamic regions are dominant. This highlights, once again, the pre-eminence of the thalamus in sleep network coordination and communication (Laufs, Walker and Lund, 2007; Schabus et al., 2007; and for a most especially useful overview see Gumenyuk et al., 2009). Thus, for decades the thalamus was considered pre-eminent in spindle generation, then the frontal lobes became briefly pre-eminent; it is now both regions in addition to key parietal and temporal areas that appear almost to kick start different kinds of slow or fast frequency spindle activity in unison during stage 2 NREM sleep. There has also been a dissociation latterly between what are called K-complexes and spindles. It is now known that K-complexes occur usually separately from spindles. Interestingly it is here there is wider agreement on function, most experts in spindle and K-complex research agree that it is these that are the actual overtures to deeper sleep (not the whole spindle) and that the primary functions of K-complexes are processing external stimuli during sleep (an evolutionary necessity) and potentially consolidating sensory and emotional memory (Caporro et al., 2012). On EEG recordings K-complexes often precede arousal wave patterns with higher frequencies. However, counter-intuitively, it seems their function is to keep the person asleep. It is partly this reason that they used to be known as "knock" patterns as they were thought entirely dependent upon a clumsy clinician knocking the patient or trace machine during sleep assessment. Although perhaps amusingly apocryphal, it is surprisingly close to the truth.

One note of caution should be sounded within this apparent consensus. That is, extremely recent research on individual differences in spindle duration, strength, and frequency also noted slightly different brain regions as paramount in their generation points. In other words, whilst generally the triad or even quad-rad of regions discussed may be very important, individual differences make these more complex in generative sites. In their groundbreaking paper Fang et al. (2019) state:

> ... EEG studies have shown that inter-individual differences in the electrophysiological characteristics of spindles (e.g. density, amplitude, duration) are highly correlated with "Reasoning" abilities (i.e. "fluid intelligence"; problem-solving skills, the ability to employ logic, identify complex patterns), but not short-term memory (STM) or verbal abilities. Spindle-dependent reactivation of brain areas recruited during new learning suggests night-to-night variations reflect offline memory processing.

In a different way, it may be argued that after almost a century since they were first discovered the true nature and extent of their functional importance is only just beginning to be understood. It appears that depending on the type of person, the personal learning preferences of the day, the intelligence (and type of intelligence) of the person all add subtleties to the geographical picture that emerges during sleep, with regard to spindle and K-complex firing, to the basic picture of the geographical regions previously described. De Gennaro et al. (2005) went as far as to suggest that sleep spindle and K-complex architecture provide an electrophysiological fingerprint due to their distinctive trait-like imprint. As has been alluded to previously, it certainly helps explain why there has been so much research that derived slightly different pictures of which brain regions were the most important for spindle generation and neuroanatomical site. If De Gennaro et al., (2005) were even partially correct regarding such fingerprints, the arguments throughout the previous century about whose research had provided the lodestone and whose research was incorrect because it used the wrong measuring devise or computer analytic tools, or some other standard criticism, seems now largely redundant.

To conclude: K-complexes and spindles are inherent properties of certain parts of NREM sleep, functionally they are most likely related to homeostatic drives (based around sensory processing, related to survival mechanisms, whilst sleeping) together with aspects of memory consolidation. Neuroanatomical regions that have consistently been shown to be generators of such activity are the frontal, temporal, parietal and bi-thalamic areas. However, the complexity and full extent of these regions appear to suggest a far more Byzantine understanding is necessary; potentially one that requires a nuanced, more individualistic approach, beyond the basic brain regions, which some argue is so complex they constitute a potential electrophysiological fingerprint.

Overall, what I have tried to do in this part of the book is to describe, analyze and explain parts of what the neuroanatomy and underlying neurotransmitter network, indeed network within systems look like, and the functions they serve. Within this, it is critical that all of us move from simplistic and rather unhelpful notions of sleep and wakefulness to a more fully worked through understanding of stages, events, and the underlying neurotransmitters that can affect these. In essence, sleep may be crudely viewed as the somnolence neurotransmitters overpowering the wakefulness neurotransmitters. However, the exact neuroanatomical geography and timing of this is incredibly complex. There are some frequently common areas of localization, but also much activity that is more widespread and the timing of different events is exquisite. The networks involved are more akin to a number of instruments playing in a vast orchestra, with numerous conductors. No single instrument and no single individual conductor is pre-eminent but they are all somehow harmonious (Figures 2.1 and 2.2).

It would be a mammoth undertaking to provide anything other than a biased and somewhat superficial view of this area. It would need an entire textbook on such topics to do it justice, and that is my essential point. I want the reader to be left with a clearer understanding of at least some of

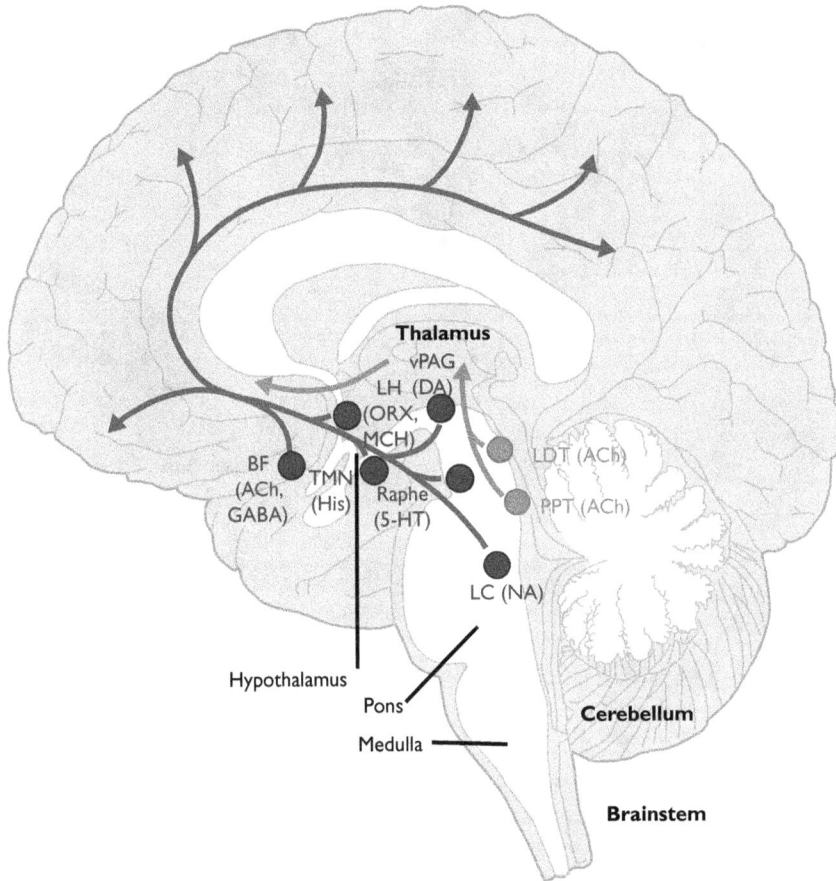

Figure 2.1 Adapted from Saper et al. (2005), wake and sleep arousal networks diagram.

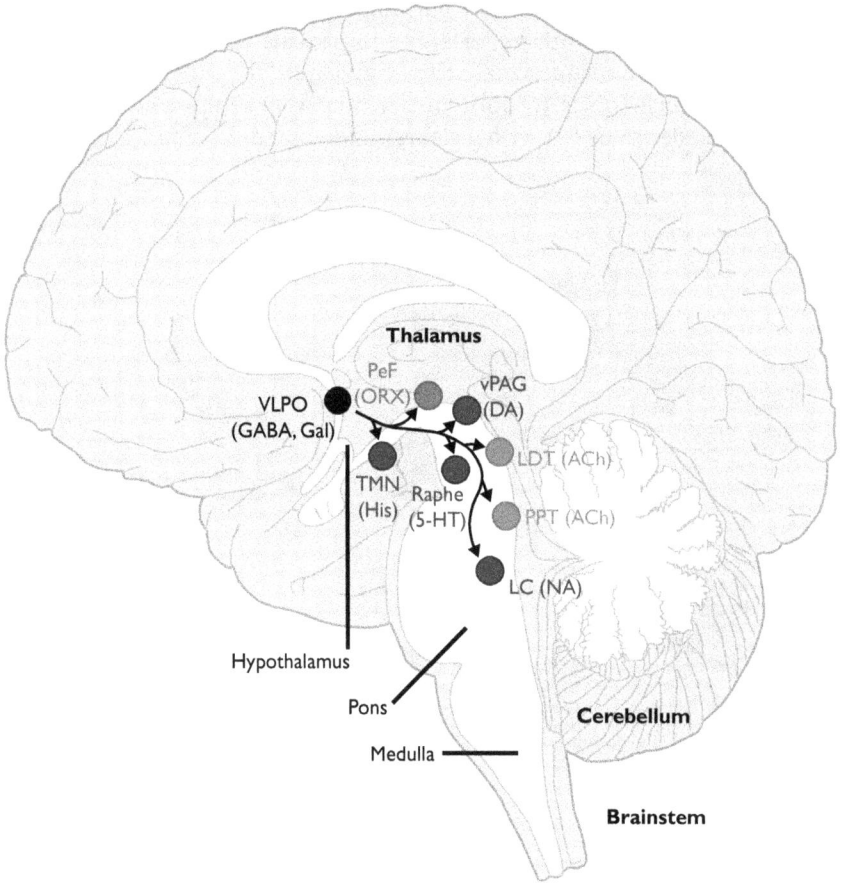

Figure 2.2 Arousal system.

this work but also potentially an excitement to learn more. For myself, understanding more about the neurotransmitters involved in sleep alone, has helped me make sense of brain injury better and that as I hope to show, will help those of us working with this population be better clinicians.

Chapter 3

What happens when things go wrong with sleep

As previously shown too much or too little sleep has significant effects on morbidity and mortality. What will now be discussed is more of the classical problems associated with sleep disorders. To begin with insomnia will be examined then sleep apnoea. The reason for mainly focusing on these conditions is two-fold. First, whole textbooks are devoted to each of these, and thus in this case brevity is key. Second, these are both intimately linked to acquired brain injuries, more than other potential classical sleep disorders, as will be discussed later. Hypersomnia is a special case, as it can overlap so easily with the chronic fatigue experienced by survivors of an acquired brain injury. In this regard, I am purposely leaving its examination and potential treatment till later in the book. That is after the keys to unlocking explanations around what goes wrong in an acquired brain injury have been presented. In any event, it is true to state that these three sleep problems (insomnia, hypersomnia, and apnoea) account for something like 70–80% of the sleep problems that occur in clinical practice, in an acquired brain injury setting.

Insomnia

Insomnia can be divided between Primary insomnia which is not linked to any other condition or issue and Secondary insomnia which is as a result of another health condition or issue (e.g. depression, cancer, or lifestyle choice such as too much alcohol). However, it is frequently virtually impossible to make this distinction, if the clinical presentation is delayed for too long after onset. Insomnia is classically considered a disorder of falling asleep, staying asleep, or early waking. This is one of the most frequent forms of sleep disorder. Even though you may have the opportunity (*sic*) to have enough sleep. Once again it becomes impossible to completely differentiate between wake and sleep issues. Having any kind of sleep disturbance in chronic form has profound effects on our daytime functioning. In nearly all cases, except facile boasting by Presidents and Primes Ministers, the majority of people

DOI: 10.4324/9780429199066-3

report feeling a worse quality of life. It impairs cognition (mainly but not solely attention, judgement, and memory functions), emotion (exacerbation of psychotic symptoms, increased anxiety and depression functions), and health (increased diabetes and other metabolic problems, cardiovascular and cerebrovascular diseases). All of us have occasional nights of impaired sleep, that is not the same, and it is here that the concept of chronicity is key. For diagnostic purposes, we should think of perhaps once a week having a sleep disturbed night the norm and not something we should fret over. Brutally it means you had a bad night's sleep, and don't have insomnia. Having more than one night a week of sleep disturbance over a longer period, say one or three months has been reported in large surveys in approximately 35% of the population (See for example, Roth, 2007). However, if some level of clear distress or impairment (as measured by a standardized scale such as The Epworth, Johns, 1991) is needed this figure falls to between 10% and 15%. The latest edition of the International Classification of Diseases (ICD-10) differentiates between organic and inorganic causes of insomnia in the first instance. The newly revised version will be fully operational in 2022, will also make this differentiation. Although the new classification is more subtle and contains more coherence with the International Classification of Sleep Disorders (ICSD – 3). The new ICD-11, describes differences in time frames, overall chronicity, and its debilitating effects. In this regard, it explicitly states that the absence of any: 'daytime impairment' will mean the person will be diagnosed as not having insomnia. It also attempts to differentiate it more clearly from other co-morbidities. Thus, chronicity as a concept is far better understood, but allows for a clinically flexible approach, unlike some older psychiatric-based reference works (e.g. DSM-V).

Causes of insomnia have previously been highlighted. In essence, they include demographic risk factors, most clearly increasing age. However, we now see post-menopausal women at greater risk than simply due to age alone. This has been hypothesized to be connected to age-related changes in the brain and hormonal system, alongside the propensity to accumulate more medical conditions and thus, medication which may harm the sleep-wake cycle (e.g. Roth, 2007). Given the dominant and wide-ranging effects of increased or decreased hormones on whole systems, this population merits specialist consideration. Other well-known risk factors are shift work; rotating shift patterns and non-24 hour cycles are especially problematic, as opposed to shift working *per se*. In addition, psychiatric and psychological problems are both causative, co-morbid and the result of sleep disruption. In an early study (Katz and McHorney, 1998,) found there was a 75–90% increased risk of at least one other co-morbid medical disorder, in those with chronic insomnia. Among those obvious co-morbid disorders are gastro-oesophageal reflux disease (GORD,) hypoxaemia, neuro-cognitive degenerative problems, and most forms of chronic pain conditions.

Treatments for insomnia

Treatments for insomnia outside of the co-morbid diagnoses, or if insomnia is secondary to one of the conditions mentioned (e.g. chronic pain), usually are a staged process from least to most invasive.

The least invasive approaches would be classified as enhanced sleep management (the awful word 'hygiene' and stimulus control procedures, examined later).

Psychoeducation regarding sleep programmes and stimulus control procedures have been demonstrated to be effective outwith any other interventions (Riemann et al., 2017). Typically, the former includes the so-called rules about good bedtime practice, including: the same wake-up time, the same bedtime time, no physical exercise for a minimum of no less than 2 hours prior to going to bed. Most especially, it includes discussion around substances. The two most prominent of these in our lives being alcohol and caffeine. The former although a central nervous system depressant, like the benzodiazepines, always produces a rebound effect during the night (some four to six hours later) which may or may not awaken the sleeper, but certainly effects the quality of the sleep. It also has a well-known reduced sensitivity over time, thus the quantity consumed has to be increased as tolerance builds up to have the same initial effect. Irrespective of the potential for alcohol dependence, this has increasingly negative effects on sleep quality and quantity. In contrast, caffeine taken in the morning prior to a lunchtime meal may be of benefit to the circadian cycle, some suggest the critical factor is the timing of ingestion. It may be one of the best available methods of helping recover from jet lag, Burke et al., (2015). This is irrespective of the supposed additional benefits for warding off Alzheimer's disease (Eskelinen and Kivipelti, 2010). In other words, a couple of good double espressos in the morning or even at lunch will cause no ill effects on sleep, indeed they may help advance our circadian rhythm slightly. This is in very high contrast with evening postprandial drinking of coffee. Most especially, the commonly held view that a coffee after an evening meal will aid digestion, whilst it may do this, it will also speedily bind to the adenosine receptors previously discussed. This process will then lead to a release of excitatory neurotransmitters, at precisely the time we should be settling down readying systems for inhibitory release and calming for actual sleep. Caffeine also has a much longer half-life than people realize. It is approximately five (5) to six (6) hours. Although, the range is very wide around this and is dependent on environmental and intrinsic factors. Hence, during pregnancy, the half-life extends from 8 to 16 hours. Whereas smoking can reduce half-life to as little as 3 hours (Temple et al., 2017).

This half-life, as has been shown repeatedly is one of the essential components in helping to understand all psychopharmacology. Crudely put this means that the example of the double espresso in the evening is better understood as containing approximately 80 mg of caffeine, after 5 to 6 hours there is still approximately 40 mg of caffeine affecting your body, as an example around midnight, if you had an early evening dinner. Even at five or six in the morning, you may still have around 20 mg of caffeine floating around your system. I say crudely as metabolism is not exactly linear and not equivalent for each person. I will not bore the reader of another pet interest of mine, the pharmacokinetics and pharmacodynamics of drugs. In other words, how drugs affect the body and how the body affects a drug. As important is the next 24 hours, by that it seems that caffeine has a longer disruptive cycle than people first imagined, if taken late in the afternoon or even worse in the evening. It would appear that the following night's sleep is also mildly disrupted by the previous evening's coffees impact on the circadian rhythm (cf. Adenosine binding). Hence its potential for use in jet lag, or even provides some support for hypersomnia, if exquisitely timed in the morning. In a study reported by O'Callaghan, Muurlink and Reid (2018) caffeine, like alcohol affected sleep, even if taken six (6) hours prior to sleep. Admittedly, this was 400 mg (equivalent to four large cups of coffee) but what was interesting, as with alcohol, none of the participants felt it affected their sleep quality, yet the polysomnography results clearly demonstrated declines in REM and NREM quality parameters, when compared to controls.

Other aspects of sleep management include modifying the environment to produce an optimal place for sleep. In effect this means, controlling light before, during, and after sleep. That means ensuring a greater focus is placed on pre-afternoon light exposure; even to the extent of exposure to blue light box therapy prior to 3 pm in the wintertime, as well as sunlight exposure in the morning throughout the year. Before sleep, it means reducing blue light exposure from modern smartphones, televisions, and artificial lighting. This is becoming increasingly difficult, as people are short-sightedly moving everything to LED-based systems, in the belief it will ease the pressure on the environment. During sleep, it means wearing eye masks, and utilizing the new blackout curtain and blinds technologies to best affect i.e. total and complete darkness during the hours of sleep. Noise abatement during sleep is seen as helpful, even though it may not be as critical as light modifications, in urban sprawls, with all the benefits of the 24-hour lifestyle, comes the drawbacks to that commonly termed light and noise pollution. If noise cannot be modified directly, then as a last resort blotting this with other forms of noise in the bedroom has been shown to have demonstrable benefits. An example is that of white noise, machines for producing such can be purchased for tens rather than hundreds of euros. However, as will be shown later their benefits are dubious, or at least more partial than we first realized (e.g. France et al., 2018; Riva, Cimino and Sanchirico, 2017).

Finally, in this part, mention should be made of temperature. As has been shown part of our homeostatic mechanism and circadian cycle is the natural rise and fall in our body temperature, which should be cool to promote sleep. This can be enhanced or interfered with by the room temperature in our house and bedroom. Often in modern centrally heated houses, our environment is too hot for optimal sleep (as well as too dry for optimal lung function). For sleep, room temperature should be approximately between 15 and 19 degrees Celsius (approximately 60 to 67 degrees Fahrenheit), anywhere below or above this range will make sleep less likely in the first instance but also of poorer quality during the night.

Stimulus control procedures hark back to the most basic learning principles outlined by famous behavioural psychologists such as Skinner in the 1950s (see for example Skinner, 1951). Stimulus control procedures if done properly are based on the deep learning achieved in childhood and earlier years (or Operant conditioning and Classical conditioning procedures). The operant principles are that contingencies are usually followed in a particular order, such as we feel sleepy we go to bed and we go to sleep, classically trained procedures can then take that forward through powerful associations of the bedroom with sleep. Other operants are that we wake at particular time windows and get out of the bedroom to go to another place for breakfast. We also do not nap in unusual places (unless we potentially have a sleep disorder); we do all of our sleeping in the bedroom. The foregoing notes on operant and classical conditioning can be seen as highly flawed in today's world. The description is perhaps more akin to most of human history up till the invention of electric light. It was at that point and the subsequent decades that things started to go very wrong for sleep. In effect, the bedroom should only have two functions; that of sleeping and sex. Nothing else should be allowed to happen in the bedroom if these associations, operants and conditions are to be sufficiently powerful to still affect sleep enough for us to be properly rested. Sadly, this is not the case, as most bedrooms have become multi-function adult playrooms, where smartphones, televisions, e-books, paper books, even gym equipment such as stationary cycles and treadmills, and other myriad distractions have been allowed to become the norm. Thus, stimulus control is most powerful, the clearer the operant or association is, with a limited set of core features. This may seem trivial to the shallow observer, but the more you can understand the underlying principles involved, the more you are likely to see the power of the approach. If we are serious about rectifying sleep and its discontents, we should be serious about all the potential approaches we know have been successful over decades. This particular behavioural approach has a vast literature backing its power in over 50 years of research and clinical practice. It should be done properly before all other approaches are tried; it is the foundation stone of good clinical sleep therapy. For an elegant meta-analytic paper, please see: Irwin, Cole and Niccasio (2006), which looked at cognitive-behavioural, relaxation, and behavioural approaches to examine their respective differences and impact on the middle-aged and over 55 population, as previously identified at risk groups.

Relaxation therapy and cognitive behaviour therapy for sleep and insomnia (known as CBT-i) is the preferred option if basic psychoeducation and stimulus control procedures fail to work, these forms of therapy will be discussed in greater detail later. Although it would appear that the behavioural aspects are the most powerful, if properly applied, even within CBTi.

A more common, and less effective, in the long-term, approach to insomnia is medication. This may at first appear as counter-intuitive, as it was the default option of many GP's, if a patient passes the brief desk assessment (often only 5–10 minutes in the UK due to pressures of patient volume), it may reasonably be assumed that its commonality means there is also a commonality of view in the medical world of inherent effectiveness of the chemical treatments used. This would be far from the reality, in the case of sleep specialists *qua* generalists or specialists in other medical or related fields. As such, a short discussion of pharmacological treatments available for insomnia will be useful at this juncture.

The two primary classes of drugs used both act on the GABA alpha subunits. These are the benzodiazepines (BZD) and the benzodiazepine receptor agonists (BzRA or the non-BZD,) or even simpler the benzo's and the Z drugs! Less frequently used drugs are those that act on melatonin receptors, those that act on orexin receptors, those that act on histamine receptors, and finally those drugs used for three other primary conditions which have proven to be helpful at inducing sleep. These latter drugs are more commonly used in depression, for the treatment of psychosis, and finally drugs used in the treatment of seizures.

First, the GABA-A The most well-known of these, and up till the last 20 years, the most commonly prescribed, are the benzodiazepines. At one point in the mid-1970s, early benzodiazepines were the most prescribed drugs in the world; they were more ubiquitous than a prawn cocktail as a starter in any restaurant. They enhance the effect of the neurotransmitter GABA. This, as was shown earlier is the most important neurotransmitter for calming, reducing muscle tone, reducing anxiety, and most critically for this part it is one of the foremost inhibitory somnolent or sedating neurotransmitters. It cannot be denied they are powerful in their effects. Indeed, in many respects part of the more recent knowledge of the side effects, tolerance, withdrawal syndromes, and negative cognitive problems occur precisely because they are so powerful. Reducing the activity of a receptor on a neuron, which means it is less likely to reach its action potential, has more than one consequence. Some of these consequences are likely to be less useful than others. Thus, the $\alpha1$ subunit receptors in the GABA-A, system of receptors are most useful for affecting sleep, the other five sub-units appear to be involved in other functions (e.g. $\alpha2$ and $\alpha3$ are mainly thought important in anxiety disorders). The benzodiazepines wash over all receptors almost equally, it is almost, as a consequence, better to view it as a wide spectrum acting agent, even though it was promoted by the pharma companies as highly specific. To be fair this is in contrast to the higher specificity of the benzodiazepine receptor agonists, although side effects are still problematic.

Partly as a consequence of the above, benzodiazepines have become less prescribed. Hopefully, even when they are they are used much more judiciously. Other than that discussed, the primary reason for this is the very long half-life of most of the benzodiazepines. Although differentiation is made between long, mid, and short-acting benzodiazepines. For example, one of the previously default option drugs in this category was Flurazepam, which has an elimination half-life of between 40 and 250 hours and it has an estimated duration of action between 48 and 120 hours. This was first produced by Roche in the late 1960s and was publicized specifically for insomnia. Although this was simply a marketing decision and not one based on clinical research (see for example: Shorter, 2005). Nearly all the so-called classical benzodiazepines have very long half-lives and durations of action. These range from Flurazepam above (along with some other early 'classical' ones) through Temazepam (T½ = 8 to 20 hours) to Oxazepam (T½ = 5 to 15 hours). Thus, even the shorter ones are not that short. What that means is they all have some effects the following day. This begs the obvious question, as to what is the purpose of insomnia intervention: to promote sleep in isolation, or to promote enough sleep, in order to function well when awake the following day? Historically the answer may have been the former, it now very clearly (as we have understood the relationship between the two states better) is the latter.

If the short-acting benzodiazepines are briefly looked at; the half-life, although undoubtedly shorter (e.g. Triazolam T½=1.5 to 5 hours, Midazolam T½=1.5 to 2.5 hours), their wide spectrum of action means a multiplication of GABA related neurotransmitter side effects. Amongst those side-effects are two critical ones: amnesia and rebound insomnia.

There are many other side effects, many of which produced the largest class-action suit in the UK ever, but we'll concentrate on the two mentioned for obvious reasons, given the title of the book. Midazolam is the most popular benzodiazepine in emergency and intensive care units across the world precisely because it is superior to most other benzodiazepines in producing impairments of memory. An example is that during its main effective dosage period endoscopy procedures are virtually never remembered, with good reason. This is not to say other benzodiazepines produce no memory deficits. Although as will be shown later, this may not always be the case depending on duration of usage and daily dosage combinations (see Mancini 2017, for a review). Other short-acting benzodiazepines (such as Triazolam; a commonly prescribed insomnia benzodiazepine) cause amnesia; indeed, it is recommended in dental sedation, precisely because it impairs memory related to the procedure. Whilst Diazepam is even better at producing amnesic effects, the downside is it has a far longer half-life (*sic*), so is not recommended simply for this reason (Parikh, 2017).

As to the rebound effect, this was dismissed early in the benzodiazepine story but gained greater weight as GP's became implicated in legal proceedings and it was discovered significant portions of the research were found to have been purposely hidden from public examination by the drug companies. Now it is uncontested, and as a result managed better. In essence, regular use of benzodiazepines produces some level of tolerance, dependency, and alterations in the GABA-A system which produce a systemic reaction when it is withdrawn. The quicker the withdrawal from chronic use the greater the reaction. Frequent, long-term users reported extensive, significant, and in some cases such negative rebound insomnia after the drugs were taken away, it was part of a cluster of reactions which produced suicidality and actual suicide. Estimates of the size of this problem are over a third of the prescribed population, and can occur in as little as 3–5 weeks of continuous usage. Unfortunately, given their potential for at least some meagre benefits described above, these kinds of significant withdrawal symptoms are most frequently found in the short-acting benzodiazepines (see for example: Neale and Smith, 2007). In their extensive review of the evidence Hintze and Edinger (2018) conclude by finding that discontinuation of these kinds of medication is: ' ... difficult to achieve.' and that whilst a gradual taper is preferred, no adjunctive medications to ease withdrawal are currently supported by good evidence. They go on to suggest that CBT-i is useful, and may even be preferred in the first instance, with a clear hint that the movement for de-prescribing also suggests benzodiazepines (indeed all hypnotics) need to be limited in usage and when they are used, they should be used for an extremely short time. I would add that given the very short time benzodiazepines take to produce withdrawal effects, and most especially in the context of this book, are they worth prescribing at all, for those with a brain injury?

Moving to the non-benzodiazepine receptor agonists (non-BzD) such as Zolpidem (T½ = 2.5 to 3.5 hours) and Zopiclone (T½ = 3.5 to 6.5 hours). There is greater hope for usefulness as they potentially have shorter half-lives and are more specific in their action, affecting primarily one subunit ($\alpha 1$), more than others in the GABA-A receptor system. The essential feature of these drugs is their benefit in reducing sleep latency and decreases in core body temperature. Thus, superficially they are attractive for those who have mild forms of anxiety specifically related to a delayed sleep onset and actual delayed onset insomnia. That aside a recent review found no clear differences between either of the Z drugs described and older benzodiazepines. However, Zolpidem may be slightly more helpful in keeping the patient asleep and not simply reducing sleep onset latency, but this statement is not unequivocal. Zolpidem also appears to help increase slow-wave sleep marginally, as well as reducing sleep onset (Wisden et al., 2017). It is also far easier to make the case that although amnesic cases and adverse reactions undoubtedly occur, they are neither frequent nor excessively debilitating as with the benzodiazepines, amounting as they do to some 1% of those taking the drugs (e.g. Tsai et al., 2009).

What is far more unequivocal is the host of contraindications, side effects, and withdrawal problems. To take one example of a contra-indication, neither drug should be prescribed to those with sleep apnoea. This is of particular concern as anywhere between 50 and 70% (depending on the study design) of those who have suffered a stroke are known to have some form of disordered breathing during sleep, the majority of which is some form of apnoea. Moreover, this disorder remains largely undiagnosed in this population. Over the last 10 years or so much debate has occurred about the exact causative nature of this relationship, prompting Alexiev and his colleagues to ask is it: 'chicken or egg?' In either direction of causality or at least association, most stroke patients are prescribed them routinely for insomnia, without any kind of sleep apnoea assessment (Alexiev et al., 2018).

Both drugs have common side effects such as headache and sinusitis. What is of more concern, is the increased association with all-cause mortality and morbidity but particularly significant increases have been shown for upper and lower respiratory infections and cancer. Whilst there is some controversy in the literature surrounding these findings the last extensive review concluded that there was very good evidence (indeed 'strong') for a: ' … causal connection between benzodiazepine and non-BZD use to motor vehicle accidents…' (at an effective dose and at least the following day), falls and fractures as a consequence of psychomotor impairment. Moreover, whilst an equally strong causal connection did not conclusively exist for much-discussed cancer, infections, and dementia, the latter investigation found: ' … the large proportion of studies concluding an association between benzodiazepines (and non-BZD's) and dementia, the criteria required to strongly substantiate a causal relationship remains only partially fulfilled.' Unquestionably, part of the reason for this is the difficulty of disentangling the issue around compromised brains and increasingly poor sleep as a dementia-like illness progresses. However, on re-reading their paper, I am reminded of the earlier tentative work of Doll and the groundbreaking epidemiological work surrounding smoking and lung disease (1950 then later in no way tentatively again in 1954,) if we were to replace strongly with moderately, there would be no debate. Moreover, using the precautionary principle so oft utilized by medicine, it should make for a pause before writing the next prescription, or at least profoundly assert their usage for the recommended maximum of two weeks (Brandt and Leong, 2017). With even stronger reason, too often the Bradford-Hill criteria that Doll and his co-author developed to examine the power of correlation are far too often ignored or remain unknown. This is very sad, as it points to a better understanding of the potential for correlation turning to causation. It is certainly true for the discussion later about the hormone, Vitamin D, which speaks to cause, rather than simple correlation.

Moreover, the non-BZD's are not superior to Temazepam (along with other BZD's in terms of incidence or severity of rebound insomnia: Voshaar, van Balkom and Zitman, 2004). Thus, even with the greater specificity no milder forms of rebound insomnia are found; the GABA-A system (and related neurotransmitter systems) will reassert itself after a period of abnormal exogenous "mismanagement", in either case.

In the earlier discussion regarding hypnotic withdrawal analyzed by Hintze and Edinger (2018) no essential differences were found between withdrawal from the non-BZD drugs and benzodiazepines, especially as is likely, the patient will have some form of rebound insomnia to cope with. The key aspects of their recommendations are try not to prescribe in the first instance, as they are not as effective as CBT-i and discontinuation is very hard. If you have to prescribe to ensure the time on these drugs is severely limited, a gradual taper is preferred after this short time. Finally (to complete the circle) CBT-i or behavioural therapy for insomnia alone improve hypnotic discontinuation outcomes.

One of the less frequently prescribed but seemingly more benign neurochemicals is synthesized melatonin (e.g. Circadin). Unsurprisingly, given the relative decline as we age of the natural endogenous form it appears to be supported for use in the elderly. As importantly, it has been demonstrated in trial after trial and meta-analysis after meta-analysis to be effective. The first large meta-analysis of note was undertaken by Buscemi et al. (2004) which concluded that: ' ... There is some evidence that melatonin is effective in treating delayed sleep phase syndrome [elderly population] with short-term use.'

Some eight years later, a meta-analysis by: Ferracioli-Oda, Qawasmi, and Bloch (2013) concluded more robustly that melatonin decreases sleep onset latency, increases total sleep time, and improves sleep quality. The effects are not as big as for the BZD's and the non-BZD's but they are important and even more importantly the side effects are far less pronounced. In 2017, Auld et al. concluded:

> Results from the meta-analysis showed the most convincing evidence for exogenous melatonin use was in reducing sleep onset latency in primary insomnia (p = 0.002), delayed sleep phase syndrome (p < 0.0001), and regulating the sleep-wake patterns in blind patients compared with placebo. These findings highlight the potential importance of melatonin in treating certain first degree sleep disorders.

This analysis was cited by the 3rd clinical update on sleep in 2018 by the Royal College of Physicians as supportive of the idea of recommending exogenous melatonin for use. However, it stopped short of fully endorsing over a quarter of a century of supporting evidence for its use. Its reticence was the relatively small number of high-quality randomized controlled trials (RCT) studies. It is not surprising that there is a lack of these, as no big pharmacological company owns patents for these. However, there are certainly dozens of good ones conducted for nearly thirty years. The evidence is better for melatonin and with fewer side effects than either BZD's or non-BZD's. For one moment let us try and find the reverse; studies that demonstrate significant contraindications, side effects, or withdrawal issues with melatonin. A search, using melatonin with the negative words listed on Pubmed, and Google Scholar provided one single citation. It produced dosage problems in young children (maximum of 4 mg up to the age of 6 years) which 'may' promote seizure activity. The older the child the greater the dose suggested was safe, up till 12 mg in adolescents. In both search databases, the majority of papers listed were how melatonin had been used to ameliorate the negative side-effects of commonly prescribed anti-psychotic medication, or other drugs and the remainder were positive accounts of its use. The old paper found was from the Lancet highlighted in a research letter that in a very small number of neurologically challenged children (four out of six) it seemed to be 'pro-convulsant' in its effects. However, every other citation under this one were reports, studies or other opinion pieces (letters, in the main) which noted its mild anti-convulsant effects. One paper examined the more recent evidence regarding glutamate and nitrous oxide reductions when exogenous melatonin was administered to children. Thus, in conclusion, a single anecdote in a front-rank journal has become poor clinical practice. Yet good quality studies, in a wide range of quality journals over decades together with many anecdotes have not become clinical practice, in the case of prescribing melatonin. The odd exception to this seems to be a more recent inclination to prescribe routinely to children diagnosed with ADHD (e.g. Masi, et al, 2019), or as previously noted its frequent usage in the over 55 population.

In another review published in 2010 that specifically looked at its safety profile in a number of different conditions and ages, Sanchez-Barcelo et al. (2010) concluded that the majority of studies examined support its use due, to the: ' … very low toxicity of melatonin over a wide range of doses.'

They also highlight how in adults it has been shown to be successfully used repeatedly in many cases for years. If the BZD or non-BZDs could have such glowing testimony, discussion of treatments for insomnia would be far simpler.

In all the studies examined that does not mean that use of melatonin is without any drawbacks or side effects, they just appear to be smaller or at the level of anecdote. For example, some describe a small 'hangover' effect the next day, which is often gone by mid-morning, if present at all. Others report headache, increases in anxiety, and some gastric disturbances as the medication is started, but these are usually reduced or gone completely after the first few weeks of usage. What is equally important to conclude in this discussion is that no paper thus far has found significant (any?) withdrawal effects after melatonin has been stopped. Moreover, for the over 55 age group it has proven to be a robust but small effect means of improving the three insomnia measures described. This effect can be further enhanced through manipulation of the suprachiasmatic nucleus (SCN) through shade wearing or blue light exposure at different times of the day depending on the age of the patient, as will be shown later.

Novel approaches to the chemical treatment of insomnia are showing some promise, but have as yet not built a consistent body of research evidence for their superiority over present methods, they are at best rather like melatonin; except far more expensive. Two of these worth considering are Suvorexant and Ramelteon. Both of these have shown some promise in clinical trials. They are particularly of interest as they do not act on the GABA-A receptors. They are orexin receptor antagonists, as has been described, this incredibly small system is the switch for either on or off of the sleep-wake systems. Orexin is the key neuromodulator within this system. The American Academy of Sleep, clinical guidelines suggest use of Suvorexant drug, even though the evidence was weak. This recommendation was based on two RCT's by the same team four years apart, funded by the drug company, which produces the drug. The sleep onset latency figures were better by 2.3 minutes for 10 mg and 22 minutes for 20 mg doses. Total sleep time on any dose was increased by an average of 10.6 minutes. Quality of sleep was not significantly different at any dose, whilst wake after sleep onset (WASO) and efficiency was significantly improved. However, the quality of the evidence for such was: 'low due to imprecision ... ' They acknowledged that the evidence was low and that that evidence wasn't what had been hoped for, later rather feebly concluding: it does appear that they do at least provide no evidence for the ' ... emergence of narcolepsy symptoms.'

As with Suvorexant, Ramelteon appears somewhat limited, but with potential. Unfortunately, at this time the effects are, again rather like melatonin, but with less of a robust and long-term research backing. Early studies appear to improve sleep onset latency by perhaps a few minutes, or 'marginal' improvement. Indeed, the latest clinical practice guidelines from the American Academy of Sleep Medicine recommend its use but with the clear proviso the evidence is 'weak'. The main reason for its recommendation was that even though the evidence was poor, it helped a little and its side effects were: 'relatively benign' (2017). Thus, sleep onset latency was improved by so little (under 10mins) that they did not feel it was clinically significant. Similarly small effects were shown for overall length (+6.58 mins), WASO (+3.5 minutes; 'well below the significance threshold of 20 mins'), quality of sleep, sleep awakenings, and sleep efficiency, showed no differences between the drug and placebo.

Overall, apart from the re-statement of the low quality of the available data, the single reason it appears for recommending Ramelteon would appear to be because it had 'relatively benign' side-effects. This would be an odd recommendation for most other drugs. The herbal remedy of standard Valerian and Hops mixture (which is 'relatively benign') available in over the counter tablets across the world had larger effects sizes in each of the categories listed (except WASO, which was not examined) and yet was not recommended later in the proceedings paper. The academicians (two of whom declared associations with Merck) recommended that most decisions should really be left to 'clinical judgement,' other than the weak recommendations for use or non-use. Their frank omission of several large and recent positive meta-analyses of melatonin amongst other odd decisions makes some of the recommendations unusual at best.

By this statement, it should not be viewed as positively biased for over-the-counter remedies, as such unquestionably, one of the most lethal drugs currently still used by vast numbers of patients across the globe, old antihistamines, bears witness. In particular, diphenhydramine short-term usage even after a few days causes daytime psychomotor retardation. Longer-term usage caused feelings of slowing and 'hangover,' particularly in the over 55 age group, and so is not recommended at all. Apart from its small to medium effect size for the actual job of reducing the symptoms of insomnia through its effects on the histamine (see earlier discussion around the wakefulness system) neurotransmitters, it is a highly potent anticholinergic. It blocks the action of acetylcholine. As we have known for some years these types of drugs should be reduced or at the very least carefully managed in the elderly (where most insomnia occurs), various scales have been developed for clinicians to administer to assess the overall anticholinergic burden. Nine such scales were analyzed by Salahudeen et al. (2015) and aside from the recommendation of particular scales one of the key factors they reiterated in their paper was the cumulative effects of such drugs on the cognitive system. Clear evidence has been demonstrated in the elderly of delirium, cognitive impairments, and finally strong associations with dementia. However, a later study by Richardson et al. (2018) appeared to provide further concerns. Rather they found good evidence of cognitive impairments on short-term usage, and anti-depressant, urological, and anti-Parkinsonian with: '... definite anticholinergic activity are linked to future dementia incidence, with associations persisting up to 20 years after exposure.' Interestingly, gastrointestinal and cardiovascular anticholinergic drugs were not shown to be associated with later dementia; suggesting something highly specific about certain classes of anticholinergic drugs. The weighting in the analysis took account of reverse causality and the potential confound of diagnosis time and post-prescription confounding. It also builds on the largest cohort study to date, that of Gray et al. (2015) in America which followed 3434 participants, prospectively for over seven years and concluded similarly about the dementia risk and strong antic-holinergic antidepressant drugs. These and the earlier study cited appear to suggest that the primary mechanism for this is the ability of certain drugs to cross the blood-brain barrior better. They were clear to differentiate selective serotonin re-uptake inhibitors (SSRI's) from this potential causal link and instead refer to older classes of antidepressants such as Amitriptyline. The problem as the knowledgeable reader will know, is that this drug has now been re-purposed as a standard drug of choice for neuropathic pain syndromes. In any event, its short term negative cognitive effects and longer-term risks for developing dementia mean, in my opinion, it should never be used in the brain-injured population, as they are already at greater risk of developing dementia in the long term and do not usually need a drug to make their cognitive problems worse in the short term.

In regards to the wider use of some anti-depressants and anti-seizure drugs, these will be analyzed later, in the final part on sleep and brain injury. Suffice it say there is as yet no particularly good evidence for their usage in primary insomnia, however, there is some evidence, depending on the drug, for their use in certain co-morbid conditions, alongside insomnia.

Amongst those primary sleep disorders that may be causative or co-morbid are: restless leg syndrome (RLS,) periodic limb movement disorders (PLMD,) and most commonly, as we shall see below, sleep apnoea.

In some respects, it can be seen that this typology can be external, internal, or a mixture. There is a third form, that of circadian rhythm disorders. These can account for some 10% of the overall insomnias. They may be secondary to losing sight and thus demonstrate a phase lag. Or indeed for those who have complete vision, including the recently discovered retinal ganglion cells (Foster et al., 1991 and on). A permanent state of jet lag occurs, which has been described as the most distressing and disabling aspect of complete blindness. Although speculated to be present since the 1920s, no real understanding occurred till much later. By retinal ganglion cells, it means those other than the more widely known light-sensing cells (rods and cones) found in the retina. This idea was initially laughed at during an American conference when it was suggested, after a few hundred years of study there may be a cell in the eye that was, until Foster and his team discovered them in mice, "unknown" (anecdote provided by Foster). Now through 30 years of supportive data across the world, we know that part of the initial processing of light before it enters the optic nerve and thence the SCN is done through these cells that are now known as intrinsically photosensitive retinal ganglion cells or ipRGC's for short (see also: Provencio et al., 2000).

Other circadian disorders include delayed and the rarer advanced sleep phase syndromes. The former often affecting "owls" with some other form of co-morbidity such as alcohol and drug use, often with the naive idea that these will help the problem, however, they often (due to addiction and rebound effects) frequently make the condition worse. The latter is often seen in: "larks" and is now understood, at least to begin with through a rare genetic disorder (Taheri and Mignot, 2002).

Sleep apnoea

Obstructive sleep apnoea (OSA,) which accounts for some 90% of all apnoeas, is defined as recurrent episodes of partial and sometimes total obstructions of the airway during sleep. Repetitive hypopnoeas and apnoeas are the result of this frequent obstruction. OSA is considered a clinical case when there are equal or greater than five (5) apnoeas and hypopnoeas per hour during sleep; the resultant figure is known as the AHI for short (the apnoea and hypopnoea index score). In most cases of classification, it is considered a syndrome when this score is combined with evidence of daytime sleepiness, often shown through scores on the ESS, or the delightfully named 'stop-bang measure.' Definitions as often occurs during first encountering a pathology have changed over the short time it has been investigated. Currently, it is generally divided into normal (<5), mild ($\geq$5–14), moderate ($\geq$15–29) and severe categories (30$\geq$).

In and of itself sleep apnoea is a debilitating condition. It causes the patient to experience numerous hypoxic events on a repeated basis nightly, usually over many years before a diagnosis is made. Apart from the obvious disturbances in breathing, sleep, and quality during the night, it has extremely deleterious effects the following day. Although a reasonable percentage of patients can have apnoea without experiencing obvious daytime sleepiness; a significant majority do experience problems the following day. It is considered a significant cause of road traffic accidents in the developed world. One study in the UK estimated it was potentially responsible for 20% of all road traffic accidents (George, 2004). In a later review, it was estimated that severe sleep apnoea, increased the risk of a road traffic accident by 123% (Gottlieb et al., 2018). Of equal concern was the conclusion which stated that: 'Sleep deficiency due to either sleep apnoea or insufficient sleep duration is strongly associated with motor vehicle crashes in the general population, independent of self-reported excessive sleepiness.'

It is this final clause that is of most concern, as it belies, as with alcohol, the idea that it is OK to drive, or at least the perception of control is greater than the actuality of control. As we discussed earlier psychomotor difficulties are one of the hallmarks of sleep deprivation and it is a particular presentation of sleep apnoea patients.

Irrespective of the conundrum of sleep apnoea causation, what is much clearer is its debilitating effects on different physiological systems. To begin with much of the earlier studies that looked at diseases of affluence and OSA left open controversies about the direction of causation. Views were entrenched about the lack of clarity, simplistic associations and the idea of chicken and egg in the whole debate of disease and or OSA linkage (see for example: Sullivan and McNamara, 1998, for a balanced review at the time). What has become clearer is that apnoea should be treated as an independent risk factor for at least some of the aforementioned diseases. Some of these diseases have much clearer directional relationships with OSA than others. In their large-scale multi-disciplinary review of the literature, Tietjens et al. (2019) described the linear, dose-dependent relationship with hypertension found in the Wisconsin Sleep Cohort. This relationship is very robust in resistant hypertension, indeed they go on to state that diagnostic testing (for OSA) is reasonable in all patients with resistant hypertension, especially as there is also an equally demonstrable reduction in systolic blood pressure in patients treated with CPAP. A number of mechanisms have been proposed for the OSA to disease linkage. One of the best supported is the repeated consequences of hypoxic episodes which occur each night. Following this the equally negative effects of repeated reperfusion injuries. This leads to acute elevations in pulmonary arterial pressures. Further chronic hypoxia activates inflammatory pathways which eventually lead to irreversible pulmonary vascular resistance increases. Although other mechanisms have been proposed in addition to these, that there is a direct causal relationship between these hypoxic events and subsequent cardio- and cerebrovascular disease-promoting endpoints is indisputable. In another recent paper, Chang et al. (2019) found that there was a direct linear positive relationship with severity of OSA and the presence of calcified carotid arterial plaques (CCAP). In this work, unlike much previous work, confounding variables for the development and severity of CCAP, including age, obesity (BMI), hypertension, diabetes mellitus, and hyperlipidaemia, were analyzed as well. When these were factored in higher AHI scores still demonstrated significant odds ratios of CCAP. This builds on much more well-known work by Marin et al. (2005), Chang, et al. (2018), and Gustafsson et al. (2018) which have clearly demonstrated the increased risk of myocardial infarction or stroke in those with CCAP. They have attempted to tease out the effects of OSA as an independent risk factor for long-term cardiovascular outcomes (again when factored for other more well-known risk factors e.g. smoking).

Finally in this regard, the recent study, which was one of the largest longitudinal studies yet undertaken, demonstrated the direction of causality; from OSA to coronary heart disease and type two diabetes, was that of Strausz, et al (2019). This used data from the FINRISK study, the Health 2000 cohort study, and the Botnia Study, based on a 25-year follow-up of almost 37,000 individuals. Its conclusion was, after allowance for potential confounding factors such as smoking, alcohol use, BMI, age, sex, HDL and total cholesterol, hypertension, type two diabetes baseline, familial history of myocardial infarction or stroke that: 'OSA is an independent risk factor for coronary heart disease, T2D, and diabetic kidney disease. This effect is more pronounced even in women, who until now have received less attention in diagnosis and treatment of OSA than men.'

Previously those suffering such disorders were seen as suffering from what many described as the: Pickwickian Syndrome, which was a reference to the adolescent character in Dickens book; *The Pickwick Papers*. Its description of the 'fat boy' is enlightening even today: *The object that presented itself to the eyes of the astonished clerk, was a boy–a wonderfully fat boy–habited as a serving lad, standing upright on the mat, with his eyes closed as if in sleep.*

Earlier, this exchange occurs:

> "Sleep!" said the old gentleman, 'he's always asleep. Goes on errands fast asleep, and snores as he waits at table." "How very odd!" said Mr. Pickwick. "Ah! odd indeed," returned the old gentleman; "I'm proud of that boy–wouldn't part with him on any account–he's a natural curiosity!"

He is further described as being very red-faced, voraciously hungry, and falling asleep, even in the middle of a supposedly enthralling or at least important task. In actuality, the boy probably suffered from a related but distinct (from OSA) condition called: Obesity Hypoventilation Syndrome (OHS).

Thus, was born both a helpful and unhelpful stereotype. Whilst it may be true to say that there is a preponderance of overweight, obese, and very obese folk who suffer from sleep apnoea. It is also true to say that significant weight gain has a far greater adverse impact on OSA than the equivalent weight loss after diagnosis, which suggests that whilst obesity plays a significant role in the development of the syndrome, it has a far less beneficial effect as a treatment. This brings us to the potential idea that sleep apnoea is a cause of obesity in itself, which then leads to a viscous cycle of further weight gain, leading to increased severity of OSA (e.g. Balachandran and Patel, 2014).

Treatments for sleep apnoea

As such if obese folk are reported (often by long-suffering spouses) to have loud snoring, they may have more likely gotten a sleep study which then confirms or discounts the diagnostic possibility of sleep apnoea. To return to the complexity issue, the stereotype has become unhelpful, the more we know about this debilitating condition. As the more we know, the more it appears that there are several factors outwith obesity that play a significant role in sleep apnoea. Indeed, the obverse of this is that somehow, personal and moral responsibility is assigned to those who don't lose weight but complain of the ongoing problem. It is now clear that for those with mild apnoea weight loss can help considerably, even cure the condition. For those with moderate OSA there is equivocal evidence of benefit from weight loss, but for severe sleep apnoea it is equally clear it has virtually no benefit. What this suggests is that within the general syndrome, there are sub-phenotypes, which respond differentially to weight loss and other interventions. There is within that clear gender differences in response. The conclusion, at this juncture, is that males with mild to moderate OSA, should be strongly encouraged to adhere to weight loss and exercise programmes (in addition to CPAP), but that those with severe OSA and women, should have different treatment schedules (see for example: Edwards et al., 2019; Carneiro-Barrera et al., 2019). This does not mean they shouldn't aim for healthy weight and fitness, as all weight loss appears to at the very least reduce the severity of OSA in some small way, rather that we should be open to the potential for less meaningful results in some phenotypes of OSA. As Edwards et al. (2019) concluded, the relationship between OSA and obesity is: ' ... non-linear and complex.' This may well lead on to the recognition that treatments such as the new cranial nerve stimulation for OSA are a suitable alternative for certain phenotypes.

In a recent study from Australia (Gray, McKenzie and Eckert, 2017) the majority of patients who were followed through a large regional sleep centre (Prince of Wales Hospital, Sydney) were either non-obese or in the normal range for weight and BMI (the latter made up 25% of the population studied). It was of particular interest to the researchers that it was also the non-obese patients who had the worst compliance with the standard treatment for sleep apnoea: continuous positive airway pressure (CPAP) machine. Their hypothesis for the reasons underlying sleep apnoea in this population was that they had lower respiratory arousal thresholds than the obese patients. A later perspective paper (Eckert, 2018) reviewed current understanding, he stated that there appeared to be at least four different, but occasionally overlapping phenotypes:

> These include a narrow, crowded, or collapsible upper airway "anatomical compromise" and "non-anatomical" contributors such as ineffective pharyngeal dilator muscle function during sleep, a low threshold for arousal to airway narrowing during sleep, and unstable control of breathing (high loop gain). Each of these phenotypes is a target for therapy.

Thus, as ever, simplistic solutions can often spring from understanding based on easy stereotypes (from Pickwick to weight loss). The reality of the condition is more complex and thus, subsequent potentially successful interventions may be multi-modal even depending on the underlying phenotype. The beguiling attraction of mythical magic bullets, once again being over-estimated.

To begin at the beginning, once again, what I have tried to demonstrate in this first part is not only the fundamentally complex nature of sleep but that understanding how fundamental it is, is really only the beginning.

Part II

Brain injury

Chapter 4

Introduction to brain injury

The basics

This chapter will discuss the basic ways different kinds of brain injury occur. It will promote an understanding of the epidemiology of acquired brain injury that is both traumatic and non-traumatic. This will be followed by the mechanism of injury or insult. For example, closed head injury from assaults and road traffic accidents (RTA's). After which the common non-traumatic brain injuries such as stroke, hypoxia, and haemorrhages. It will point to the fact that outside neonatal care these patients are the most vulnerable group healthcare services worldwide tend to. Its purpose is to draw threads together which may be utilized to better effect for clinical benefit, than is currently the case. My intention is to take a quick deep dive into the secondary effects of these injuries, rather than, as with many texts, concentrate on the initial phase of these injuries.

Scale of the problem

To begin with it is important to gauge the magnitude of the problem. Acquired brain injury in its broadest sense is any injury to the brain which occurs after birth. The kinds of injuries include the more obvious non-traumatic ones of stroke and tumours, together with the more obvious traumatic injuries that occur as a consequence of a RTA or an assault. In addition, they include less well-known acquired brain injuries such as: hypoxia, encephalitis, meningitis, and auto-immune problems. Practitioners of Neuropsychology and campaigning organizations in the UK, such as Headway, are concerned with the three primary consequences of these brain injuries. These are cognitive (e.g. memory, executive disorders), behavioural (disinhibition, verbal outbursts), and emotional (e.g. anxiety, depression). In addition to the more obvious physical problems, that are the province of others. According to the most recent government instigated report, co-produced by the United Kingdom Association of Brain Injury Forum (UKABIF, 2018) between 293,000 and 301,500 people suffer an acquired brain injury each year in England (predominantly stroke and traumatic head injury). To gain some perspective on these figures, according to the ONS

DOI: 10.4324/9780429199066-4

(gov.data.2017) there were 305,683 new cancer diagnoses in 2017. As is well known, although stroke incidence has reduced over the last few years it still remains the third highest cause of mortality in the UK. These figures are largely similar across developed nations around the world. For example, it is the fifth-highest cause in the USA (centers for disease control; CDC, 2016) and in Australia it is the third. The problem of morbidity and mortality from acquired brain injury is thus huge, in all developed countries. According to Kamalakannan et al. (2015) it is described in nearly identical terms in India to developed countries, they suggest it is a: 'silent epidemic.' They go on to review the epidemiological data and demonstrate an exponential increase of incidence figures that are sadly catching up with other countries, as Indian development accelerates.

Mechanisms of injury

Some thoughts on the nature of traumatic injuries, to begin with, would reasonably begin by a better understanding of the commonest; closed head injury or traumatic brain injury (TBI).

Through many centuries, but particularly the last two, enormous developments in understanding of the biomechanics of TBI have occurred. Although producing sometimes contrasting results animal, computational, strain studies in primates and humans, fMRI and magneto-encephalography (MEG) studies, cadaver studies have all contributed to this deeper understanding, when they have been pieced together.

There are two primary mechanisms of TBI, these are the direct force of the initial impact, with whatever force makes contact with the skull. This is then translated into an impulse or force which causes indirect rotational and linear effects. These mechanisms produce increased intracranial pressure. However, it has been extrapolated principally from primate experiments, human studies, and computational models that the rotational forces cause the greatest damage in TBI. Indeed it has been estimated that 90% of the damage is due to these rotational forces. Using the same multi-data products, it has been argued that forces acting upon the coronal plane (front to back) produce the worst damage and thus clinical outcomes, although repeated twisting forces, in which the coronal plane dominate, in terms of the direction of force, such that can be found primarily in boxing, produces even greater damage over time even though consciousness is intact (see for example: Baird et al., 2010).

After these forces have brought about the first waves of havoc, the neurochemical, cellular, and inter-cellular damage begins. This biochemical cascade is longer lasting than the more immediately understandable physical damage. Unusually, given the nature of the basic sciences, understanding is easier for the primary science. In this regard, the Physics of TBI is easier to comprehend than the Biochemistry of TBI, indeed the latter is still not fully understood (Prins et al., 2013). Broadly these processes have been conceptualized as oxidative stress, multiple electrolyte disturbances, glucose metabolism changes, and resultant neuroinflammatory processes. The latter being one of the important themes of this book.

The initial injury described above causes massive neuronal depolarization and firing. Glutamate is released between cells. This excitatory neurotransmitter is the most common one in the brain, it has been estimated to be involved in over half of all synaptic action potentials (see for example: Purves et al., 2011). Glutamate is critical for information processing and essential fast signalling. There are widespread glutamate receptors found throughout the cortex, very densely in the hippocampi, even in glial cells. It is inextricably linked to nearly all other neurotransmitter systems, along with aspartate. In his succinct introduction to a small part of this and other systems Kondziella (2017), notes the right amount needs releasing. Indeed, it has been known for some while that very specific amounts need to be released in exactly the right places for a very small amount of time. Too little slows inter-neuronal communication, whereas too much is basically neurotoxic (Meldrum, 2000, and later Guerriero, Giza and Rotenberg, 2015). In brain injury massive amounts are released and in the wrong places. This inappropriate release of glutamate, amongst the more direct neurotoxic effects also causes greater potassium release, which as the primary determinant of resting membrane potential interferes with the threshold potential. The combined effect is to propel massive unregulated neuronal depolarization and firing causing further release of excitatory neurotransmitters and their precursors, which then ultimately leads to greater potassium imbalances, and so it goes on. A number of consequences of these processes have been found. In mouse experiments controlled blunt force injury has led to epileptiform discharges. Whilst this may be viewed as an obvious consequence of glutamate flooding, in humans the counter-intuitive slowing of background frequency of EEG waves has been found; an effect that lasts many months (Nuwer et al., 2005; Slobounov et al., 2012).

Following on from these injuries and problems, the brains healing mechanisms start to kick in. However, they cannot cope with the quantity of imbalance. Glial compensation is not adequate. There is then an increase in enzyme activity which partly (solely?) propels increases in energy expenditure, the exact degree of increased metabolic demand is not being entirely clear in humans. It is understood to be greater in the first few weeks after such injuries. Alongside this time frame, there are parallel increases in intra and extra-cellular calcium. This process leads to impaired oxidative metabolism and has been hypothesized to be integral to neuronal cell death (Weber, 2012). Shifts in magnesium at the intracellular level also occur after TBI. This significant depletion is also linked to unregulated cell apoptosis. This is important, as although apoptosis is frequently occurring; it occurs in a regulated and planned way to clear away old cells and is part of the process of renewal. As such magnesium was touted as a potential supplement after a TBI, however, no trials to date have provided clear supporting evidence for its use, indeed in at least one large trial it caused poorer outcomes (Temkin et al., 2007). However, in a well-thought-out review Carbonara et al. (2018) have argued these mixed results are primarily to do with the time difficulty of finding the optimal temporal frame for administration. Time and the timing of various processes and subsequent supports is another theme in this book.

The commonly understood neurovascular coupling found in normal brain functioning, in essence suggests a beautifully regulated and inextricably bound system with metabolic demand on one side of the coin and cerebral blood flow on the other. This relationship, or at least its tightly bound nature is lost after a TBI. As the increased energy demand needed through increased potentiation and over-production of glutamate at its heart occurs, cerebral blood flow cannot cope with demand. At a time of even greater oxygen and other nutrient demand (especially glucose) the damaged brain is less capable of optimal delivery (Bondi et al., 2015).

At its most basic energy-producing level mitochondrial impairments limit adenosine triphosphate (ATP) synthesis in neurons. Compensation occurs but the limitations of these sub-standard energy products limit cells damaged in a TBI (that have survived) to perhaps only around 10% of previous production. This has been partly put down to cell damage but also various oxidative metabolic problems post-injury. The metabolic process in the early hours and days of recovery have been shown to go from extreme hypermetabolic states to the polar opposite; hypometabolic states. Moreover, the greater the decrease in glucose metabolism in grey matter in the first days of injury, the worse the outcome at 12 months (Wu et al., 2004; McGinn and Povlishock, 2016). It is in these setting events that a metabolic crisis occurs, during which time erratic energy demands at the cellular level are met with poor to non-existent supply issues.

Two common supposed non-traumatic acquired brain injuries

Hypoxic brain injury is one of the more common non-traumatic brain injuries; especially as a result of cardiac arrest. Not only a leading cause of mortality it has become an increasingly problematic cause of chronic neuro-disability; as trauma care has improved, more cases of living with long-term problems have increased. The immediate reduction, or in some cases complete cessation of the supply of oxygen is followed quickly by a secondary injurious cascade, ultimately leading to further neuronal cell injury and cell death (for an overview see Busl et al., 2010).

The incidence of cardiac arrest is around 30,000 per year in the UK. These occur mainly out of hospital and less than one in ten survive even after being admitted to hospital. At least 5% will have some kind of permanent neurological damage to cope with and although there is wide variability in outcome and earlier ambulance resuscitation data, the overall survival rates are very poor (Perkins et al, OHCAO Project Group, 2015). The hypoxic injury causes neurological disability which often range from mild attentional difficulties through to persistent vegetative states. In addition, for those who do survive there are frequently considerable co-morbidities, such as anxiety, depression as well as more obvious physical problems (Nolan et al., 2008). Despite huge efforts, and in contrast to many other critical conditions, hypoxic injury from such events has not improved significantly in over 20 years. That is apart from one single intervention, which has shown good effect, that of hypothermic management after the arrest (Nielsen et al., 2013).

Internally, cerebral oxygen depletion occurs rapidly during a cardiac arrest. As the brain is such a hungry organ without any nutritional storage a crisis in blood supply leads within minutes to neuroglycopenia and cell death. From what we know through experimental models and clinical observation the magnitude of ischaemic tissue loss and the severity of neurological symptoms are related to the length of cardiac arrest and the total downtime period. Unfortunately, this is virtually always vaguely known unless dispassionately witnessed. More recently, evidence has accumulated which demonstrates further support for the supposed two-hit hypothesis regarding hypoxic injuries and the final extent of brain injury. Sekhon, Ainslie and Griesdale (2017) discuss the clinical pathophysiology of hypoxic injury and note that the second hit potentially can be seen in terms of the classical understanding of the problems inherent in reperfusion. In that the re-introduction of oxygen itself although viewed by the layperson as inherently good, will actually cause harm, in this instance. By this, it is understood to be the linked mechanisms of loss of ATP production aerobically, via a shortage or complete absence of oxygen. Although this was glanced over previously when discussing the secondary injury in TBI, it needs further examination. In this regard, a brief summary of the basic biochemistry is needed through an understanding of the electron transport chain. This chain is in essence a hydrogen pressure reactor (or at least mill turned by hydrogen) which by the end of the process pumps out around 32 ATP units. These are necessary for the sodium and potassium pumps regulating both, in all cells of the body. This chain has five components or compartments; compartment four (4) is entirely dependent on the delivery of oxygen through red blood cells. Now if through cardiac arrest or similar problem, oxygen is not delivered to red blood cells (RBCs) via the lungs, they cannot subsequently deliver that to compartment four and the ATP production falls or eventually stops completely. A number of important processes then kick in. The first is like a "back-up generator," which relies on less efficient anaerobic production, starts to work. However, this is inherently inefficient. It only produces, within equivalent parameters, two-three (2–3) ATP units. As production continues a by-product of this anaerobic process is lactic acid, as this lactic acid builds in cells, it leads to lactic acidosis (similar to the process involved in muscle cramps after significant exercise); which begins to harm the cell. In addition, as ATP production diminishes the classical three sodium molecules out and two potassium molecules in (the pump) begins to fail leading to too much sodium in the cell and too much potassium outside. As water is attracted to sodium the cells start to suck in larger and larger quantities until each expands beyond its naturally elastic capacity causing cell wall damage. This is combined with the damage caused by the increased acid. Moreover, the calcium channel becomes disturbed which also leads to direct cell damage and mitochondrial toxicity, leading to parallel further disturbances in ATP. This whole process is what may be described as one aspect of the first hit. It is the consequence of oxygen depletion and eventual complete starvation, leading to cell ischaemia and finally necrosis.

So far so simple, however, we now know that subsequent re-introduction to damaged cells that have not been completely destroyed during the initial phase of oxygen starvation leads to the second hit. At least four processes have been implicated in this hit. First, it is the oxygen itself that causes harm. This is primarily due to the depletion of anti-oxidants as the cell begins to become harmed, oxygen is then introduced to a place where there are many free-floating reactive oxygen species without breaks and buffers to react against the oxygen itself; producing further significant harms. Second, pro-inflammatory responses occur, such as certain cytokine processes and inappropriate glutamate release, inducing necrosis or unplanned apoptosis. To be clear, a cytokine (from the Greek for cell and movement), is simply a small protein messenger, there are many and they are important for proper health; especially our immune responses, but as with so many processes, they can go wrong. In this sense, and to recap, apoptosis classically understood is without inflammation, it is a planned process. Third, vasomotor constriction and other vascular abnormalities potentially, but not exclusively, caused by decreases in nitrous oxide production. Finally, mechanistic explanations based on resultant localized and para-local and generalized oedema, from cell, to cerebrovascular endothelium to larger tissue to wider organ structures. This has now been further elucidated by Sekhon et al. (2017).

What is more commonly understood over many decades of research is that it only takes 15 minutes for over 90% of brain tissue to be damaged in some way, and that this damage begins to take place after only 30 seconds (Ames et al., 1968; Shemie and Gardiner, 2018). This is partly why sleep apnoea has the potential for causing a chronic form of hypoxic injury, some of which may not be recoverable from; in a dose (severity) and time (usually years) dependent manner akin to the calculations on the damage to a smoker in pack-years (World Health Organisation; WHO, 2008). What has not been studied adequately is the potential for repeated low-level reperfusion injuries, as a consequence of the repeated hypoxic injuries. When it has been discussed it is usually as an adjunct to another focus (Jaster, 2018).

Stroke or cerebrovascular accident (CVA) is common, so common that it stands as the third or fourth-largest killer in developed countries. In the UK according to the Stroke Association (2017): 'Age standardised stroke incidence was 115 per 100,000 people. There are more than 100,000 strokes in the UK each year. That is around one stroke every five minutes in the UK.'

Overall, worldwide stroke is the second principal cause of mortality after ischaemic heart disease. Although stroke has not received the much wider publicity of other killers. So many known and somewhat erroneous gender-based assumptions around ill-health could be slain at this juncture; I'll simply take one. A woman is 50% less likely to die from breast cancer than stroke and a man is twice as likely to die from a stroke as he is from prostate and testicular cancer combined.

Depending on the type, site, and severity of the stroke, long-term neuro-disability can range, as examples, from very mild speech problems (for a small discreet dominant cortical stroke) through persistent vegetative states (for a large severe grade haemorrhagic stroke). In a large study reporting over a decade, using the South London Stroke Register, significant problems were discovered in extended follow-up. In their main conclusion, Wolfe et al. (2011), found: 'Between 20 and 30% of stroke survivors have a poor range of outcomes up to 10 yrs after stroke.' Unsurprisingly, age was associated with higher rates of disability, inactivity, and cognitive impairment. This paper also examined the most common forms of unmet needs and offered explanations based on environmental factors outwith age and severity of stroke. The primary findings were that psychosocial, cognitive, fatigue, and pain were most prominent. In addition, leisure time and employment problems were prominent in the wider domain. So much rehabilitation efforts are placed on physical and related problems, resulting in mobility being the only need actually met identified longer-term, causing at least a pause in the programs currently available.

In regards, to specific clinical pathophysiology, what is increasingly apparent are the similarities between stroke pathophysiology and hypoxic injury previously discussed. In the first instance, an ischaemic event such as thrombosis or an embolus (haemorrhagic strokes are outlined later) leads to immediate hypoxic injury at the site of the event. The intermediate stage of potential harm is understood through an area known as the penumbra. This area surrounds the specific site of the ischaemic event, it is tissue that is not yet irreversibly damaged but has the potential to become so.

In other words, it may be supported through damaged collateral arteries that supply the penumbral area and are still viable, even though potentially damaged. One of the key clinical discussions during the early hours after a stroke, through blockage, is how best to protect this area from full damage, in this sense, ensuring the oxygen levels to cells in this area are sustained or at the very least not completely lost. As the hours progress penumbral tissue deteriorates and gradually becomes incorporated into the irreversibly ischaemic core of damage.

Guadagno et al. (2003) discuss the important advances outside the gold standard of positron emission tomography (PET) scans to understand at least the visualization process. Diffusion and perfusion-weighted scans now offer rapid assessment of the critical penumbral area. This is essential as they attest to the direct impact of penumbral size and speed of response to final neurological outcome. These kinds of more recent techniques alongside others have become critical in improving outcomes. This is primarily due to improvements in acute diagnostic scanning which have led to swifter decisions about thrombolytic treatments and as we know, the earlier the onset to treatment time the greater the benefit and overall prospect of good recovery (see for example: Rother, et al., 2013). Unfortunately, treatments, in this case, have not kept pace with the improvements in diagnostic measures. Henninger and Fisher (2016) estimated that less than 10% of ischaemic stroke patients receive intravenous thrombolysis; they go on to estimate that this leaves a further population almost twice that size who could benefit from such treatment. In addition, they are keen to point out that leaves the vast majority of cases where these treatments are not appropriate, potentially due to the risk of subsequent haemorrhage due to the treatment itself. There is thus huge scope for the development of therapies where anti-thrombolytic (through drugs or mechanical means) therapies are contraindicated.

As time has moved on, critical features of the ischaemic cascade have been further identified. As was described earlier, in hypoxic injury, an important feature in this cascade is inflammatory processes. Primarily these are shown through early prominence of cytokines especially the subsidiary form: chemokines. The former has been extensively researched recently. The cascade previously described during oxygen depletion which leads to ischaemia and the subsequent further cascade on reperfusion of oxygen leads to the breakdown (amongst other problems highlighted during the discussion on hypoxic injury) to the breach of the blood-brain barrier. This allows immune cells such as T-cells and neutrophils to enter and accumulate in the damaged tissue. Alongside this, microglia (glial cell macrophages resident in the brain and spinal cord) become strongly activated due to increase in ATP and other inappropriately positioned matter such as glutamate outside the cell. The microglia secrete cytokines. On one level this activation can be helpful, as neurotrophic factor is produced and dead tissue is cleared away, along with other potentially harmful waste. However, the self-same activation causes pro-inflammatory cytokines such as tumour necrosis factor-α (TNF-α) to be released alongside yet more reactive oxygen species. A further cascade results as ischaemia and unregulated cell death occurs, producing even more inflammatory signalling. A feed-forward response occurs which ultimately leads to greater cell death in the acute phases. The level of these inflammatory changes has been strongly implicated in the extent of brain damage during cerebral ischaemia; specifically, the increase in pro- as opposed to anti-inflammatory cytokines (Doll et al., 2014). Of critical importance for later discussion are the post-mortem brain tissue studies undertaken in humans. TNF-α, positive cells are found in all ischaemic tissue three days after stroke, moreover they have been found up to 15 months post-stroke; the majority being microglia and macrophages. As further research has progressed, especially in the last five years, our knowledge of timing and effect has become more nuanced. In this sense, inflammatory processes are now seen as critical for both extent and degree of damage and thus neurological outcome. However, at different times these processes are also crucial for the degree of subsequent recovery. What has also become better understood is the greater length of each of these harmful or helpful processes. Initially, these were thought critical in the minutes and hours following a stroke (or indeed other brain injuries), now these time-frames have become further extended to days, weeks, and now months. It would appear that different waves of pro- and anti-inflammatory processes occur at different stages of injury and then recovery. An example is blood-derived (as opposed to resident) macrophages are largely absent in the first few days after ischaemic injury but then gather pace to peak after the seventh day, after which they decline. Neutrophils on the other hand infiltrate the ischaemic tissue within the first 30 minutes then peak between one and three days. T lymphocytes whilst largely absent in the early minutes after damage has occurred, gradually increase between days three and seven (Brait et al., 2012, Yoshimura and Minako, 2020). At least some of these processes have been clustered under the recent heading of damage-associated molecular patterns (DAMPS) and are particularly prominent in the first weeks and months post-stroke. In a separate but related stream of research post-stroke inflammation has been related to depression, although the results here are not entirely unidirectional (Pascoe et al., 2017). Of greater relevance to the current chapter is the more well-known and robust evidence for long-term inflammation to be present when compared with matched controls and for a preponderance of pro-inflammatory phenotypes in chronic neurodegenerative conditions such as multiple sclerosis (MS) and Alzheimer's (Zindler and Zipp, 2010). Indeed, it has been proposed, the higher the inflammation the more likely the condition deteriorates faster (e.g. Fang et al., 2019; Noonan et al., 2013; Rajkovic et al., 2018). Much research efforts have been undertaken to find the best time and the best way to modulate these pro-inflammatory responses in the early hours and days. Although, as I have suggested this may not actually be such a good idea, as so many of these inflammatory processes are actually useful. However, as understanding has become more nuanced, similar efforts have not been made to tackle the chronic forms of these pro-inflammatory problems. It is these that would make an enormous difference to the long-term outcomes of acquired brain injury of all kinds. In some respects, chronic poorly regulated inflammation is the final common pathway of any brain injury course and long-term outcome. As such research and clinical efforts to ameliorate these problems have been woefully lacking. Having said this research has begun to be taken more seriously, as even a dedicated open-access journal was instigated in 2004, called: *The Journal of Neuroinflammation.*

Interesting recent research by Younger and colleagues (2017) initially demonstrating clear elevations in TBI and ME/cfs patients of lactate (see above). More recent research by the same team, found in addition to this, even higher levels of *myo*-inositol for the TBI sub-group (Mueller et al., 2020). In effect, this is a marker of increased microglial inflammatory activity, as opposed to other cells. It suggests that in the areas identified too much microglial and thus chronic inflammation was occurring. In other parallel research regional brain temperature has been utilized to measure neuroinflammation by proxy. Karaszewski et al. (2013), found this to be the case in a sample of patients who had suffered an ischaemic stroke when compared to controls. Moreover, they also suggest that the higher the temperature the worse the outcome at follow-up. By this we come back to the same problems of the difference between acute and chronic events. In this sense there is always an increase in localized temperature associated with the discreet brain injury event. However, this recedes over time, except in those with poorer outcomes. In addition, whilst increased inflammation and thus temperature is "normal" in the acute phase; it is the degree of temperature increase that is of central importance. Not only this but the areas found that were producing the most lactate (e.g. cingulate, hypothalamus, and hippocampi) were also found to be the ones that had the highest temperatures. As we know these areas are profoundly important for neuronal recovery. Karaszewski et al. (2013), found that when the temperature moves from local to a systemic pyrexia, then even worse outcomes occur. The importance of the principles involved in temperature regulation, inflammation, and outcome will be returned to later.

Whilst there are obvious shared aspects of both acute and later, even chronic biochemical changes that occur in all the previously discussed brain injuries, there are also extra highly specific pathophysiological changes to do with haemorrhagic strokes. Most clearly this can be understood through the large amounts of extracellular blood and its contents. As intracerebral haemorrhage (ICH) is the most common type of haemorrhagic stroke, accounting for between 10% and 20% of all strokes (Ikram et al., 2012). This will be the focus of what follows. It is a particularly severe form of stroke, with a high mortality rate (between 30% and 50%) and a high morbidity, with 74% of those who survive being functionally dependent a year after the event (van Asch et al., 2010). Increasing population age and more frequent anti-coagulant usage may increase this form of stroke. Broderick (2020) demonstrated the most frequently found cause is some form of hypertension, especially as demonstrated through hypertensive microangiopathy found in deeper structures such as the basal ganglia and brainstem (slightly less frequently in the cerebellum).

Apart from the Glasgow Coma Scale (GCS) and Post-traumatic Amnesia (PTA), the best predictors of outcome are the volume of the haemorrhage (larger than 30 cc), age (older than 80 years), and depth (lower in the brain nearer the brainstem) the worse the outcome. The more obvious mechanical deleterious effects of such haemorrhages are best understood through the mechanisms previously described above for TBI and later hypoxic brain injuries (including mass effect, pure hypoxic injury, and subsequent oedema: cellular, tissue, and wider systems). Indeed, even the later discussion is relevant here as over 30% of ICH patients suffer at least one ischaemic event within the following month (Menon et al., 2012).

The major components of secondary brain injury in this regard are haemin and iron which have previously been converted from the free-floating haemoglobin, these two components are very neurotoxic. In other words, the degraded products partly destroyed through the breakdown of blood products in areas outside the cerebrovascular system cause further harm to healthy brain tissue. They result in direct neuronal damage and localized oedema (Huang et al., 2002). One of the primary reasons for such neurotoxicity is oxidative stress and iron-instigated production of free radicals via a process called the Fenton-reaction (Clark et al., 2008). This complex reaction is, in crude summary, when iron is in the presence of hydrogen peroxide it produces hydroxyl radicals. These particular reactive oxygen species (ROS) have been implicated in the pathophysiological mechanisms underlying Alzheimer's disease, with both neuroinflammatory and neurodegenerative effects (Kanti-Das et al., 2015).

Microglia resident in the brain are activated within minutes through signalling of direct tissue damage. They are also specifically activated blood components such as thrombin and haemin that trigger inflammatory responses, which contribute to initial neuronal damage. Both microglia and microphages activated have either of two phenotypes (M1 and M2). M1 produces an abundance of pro-inflammatory reactions, with resultant destructive cytokines. In the early phases of recovery the M1 microglia promote further neuroinflammation through the additional burden of the blood-derived substances it is helping to clear in the extracellular spaces (e.g. fibrin). However, M2 has been associated with neuroprotection and regeneration after brain injury, once again pointing to the complex nature of inflammatory processes in brain injury (Zhang et al., 2014). It has been postulated that the reasons for the poor clinical trials of anti-inflammatory drugs are because of the complex bi-directional nature of inflammation processes and critically the time when these have been administered (Mracsko and Veltkamp, 2014). An example of a negative pro-inflammatory reaction to intra-cerebral blood is that these processes result in vasoconstriction of local arterioles and capillaries, which have been proposed as one of the central mechanisms in subsequent ischaemic strokes after an ICH. A contrary effect of speedy clearance of intra-cerebral blood (e.g. surgical removal of clots) is that this process may limit the chemokines necessary for triggering M2 microglia, which as has been stated are critical to some aspects of recovery in ICH (Taylor and Sansing, 2013). In other words, microglial function is dependent on time since event but as shown earlier in other recent work, M1 produces *myo*-inositol which is a marker of the level of neuroinflammation and subsequent increases in temperature, which it appears then can become in certain brain injuries a chronic cycle of neuroinflammatory responses akin to ME/cfs conditions.

Similar complex inflammatory functions dependent upon time have been found for certain leucocytes, for example neutrophils. These are the first leucocytes to migrate into the injured brain. Later after peaking around 3–5 days after the event, they begin to be important for neurogenesis and subsequent outcomes (e.g. Tsuyama et al., 2018). This is crucial to later discussion, inflammation is not an either or, as can be simplistically found in the self-help sections in book shops. Even more importantly; it is more or less helpful, for different places in the body, at different *time* points.

The purpose of the aforementioned discussion was to illustrate the complexity of the problem at hand, it was not intended to be a full examination, but a primer for what is to follow, for a more comprehensive review see for example: Mracsko and Veltkamp (2014) or for an enlightening and related study of the differences and similarities of the ICH versus the ischaemic transcriptome (which points to the earlier work discussed) see Stamova et al. (2019). A number of potential ideas and conclusions may be drawn from this. First, there are both good and bad inflammatory processes, that occur at different times. Many of the earlier inflammatory processes are crucial for recovery, indeed some aspects, as will be shown, of neuroplasticity are almost entirely dependent upon the signals from early inflammatory processes. However, some of these inflammatory processes are less helpful and potentially go on longer than necessary. Moreover, others that are either the result of these earlier processes or which simply continue at a lower level of "grumble" become chronic features of recovery from a wide range of brain injuries. At these later time points, chronic inflammatory processes become entirely unhelpful and cause almost an infinite perpetual motion of inappropriate signalling and then inflammatory release; as the system itself becomes problematic and self-perpetuating. In some respects, these processes at their worst become more closely resembling an auto-immune-related disorder.

As will be demonstrated later, the precise interventions which may alleviate some of these chronic pro-inflammatory responses may have been beside us all along.

Chapter 5

Common neuroanatomical structures that are affected in brain injury

As was described earlier TBI's, hypoxic brain injuries, and strokes have specific and well-known neuropathological and neuroanatomical features. This chapter moves us from the bio-chemical, neuronal, and extracellular to these more well-known higher-order features.

In order to drill down and act as a mirror for the equivalent chapter in Part 1 (Ch.2), the neuroanatomical structures most readily damaged will be discussed. Thus, principally the frontal and temporal lobes have been most frequently damaged in brain injuries associated with RTA's. These areas have been known to be the focus of contusions since Courville's ground-breaking work examining post-mortem remains of traumatic deaths. In contradistinction, in hypoxic cases unusual features stemming from the Circle of Willis, frequently effect the base connections between the temporal and frontal lobes, but even more typically the hippocampi, basal ganglia, and cerebellum.

TBI

TBI's have by definition some form of traumatic insult. This does not ne-cessarily mean the brain has to be struck by something, it can result from a high-speed acceleration than a deceleration incident. Perhaps this was best illustrated by Gennarelli and colleagues in primate models during the 1970s and 1980s. Today it is more widely accepted that some form of brain injury can readily occur in the absence of a direct blow of some description. What occurs if the acceleration and deceleration are fast enough, with or without an initial blow, are very similar patterns of neuroanatomical pathology (e.g. Barth et al., 2001). In particular, it is well known that the frontal and temporal lobes are vulnerable to such trauma.

In large measure, an understanding of the cranial fossae (any of the three base cranial or skull structures), will inform how the soft tissue of the brain is impacted through a TBI. In normal life, these bone structures hold the soft brain in place well, with an allowance for some limited buoyancy due to the cerebrospinal fluid, and the meninges, between the skull and the brain itself.

DOI: 10.4324/9780429199066-5

It is this combination of smoothness of some layers and the uneven contours of others that ordinarily allows the brain to stay held and not slosh around too much but also to move slightly when needed. Other features which hold the brain gently in place are the falx cerebri, running down the middle between the two hemispheres, and the tentorium cerebelli found between the cerebellum and primarily the occipital lobe, with some aspects of the temporal and parietal regions also above it. For the most part of human activities, such as running, jumping, play and real fighting, even falling or moving and then halting at reasonable speed, these structures are highly protective. It was obviously a great advantage in evolutionary terms. If we had not evolved in this way survival would have been harder, to hunt prey or indeed run as the prey, was critical to our success as a species.

Unfortunately, high-velocity movement and then deceleration are not something this design takes account of. The extremes of movement most apparent from the 20th century onwards have outstripped our basic evolution. Courville himself was partly motivated into better understanding the properties of neuropathology associated with the increasingly common deaths from the then recently invented car (1937). Speed, indeed hyper-speed, in all things is the biggest factor in human development in the last 100 years. However, this has led to distinct disadvantages when it comes to brain injury, as the very structures that gently but firmly hold the brain normally are responsible in large measure for the worst aspects of the initial brain injury.

One specific aspect of this "holding" fossae was of particular interest to Courville. The sphenoid ridge is nearer the anterior part of the base of the skull. It runs laterally along the underside of the anterior temporal and inferior to posterior aspects of the frontal lobe. In his series tracing of injuries over many post-mortem examinations he noticed that these regions were the most likely to have sustained contusions, see Figure 5.1.

Whilst later more sensitive scan studies have found a domino effect (and atrophy) on the corpus callosum (CC) and most especially the fibres attached to the CC from the frontal and temporal regions, these studies, have also largely confirmed this earlier groundbreaking work (see for example: Wu et al., 2010).

Early neuropathologists described the action of these and other bone protuberances against the brain as a "slapping" action. The brain briefly is lifted then slapped against these structures in high-velocity injuries. As has been outlined these forces make for contusions at the sites of the slap but also additional shearing and tearing forces on the axon fibres attached to them, and elsewhere, if the force is high enough. Not only this, these forces obviously result in shearing and tearing forces against blood vessels, leading to the ischaemic and haemorrhagic changes previously analyzed (Marion et al., 2018). This is further complicated by the physics of movement including acceleration and deceleration, which mean a variety of differing forces exert influence on the final primary brain damage (e.g. direct linear impact, oblique rotational).

(a)

(c)

(b)

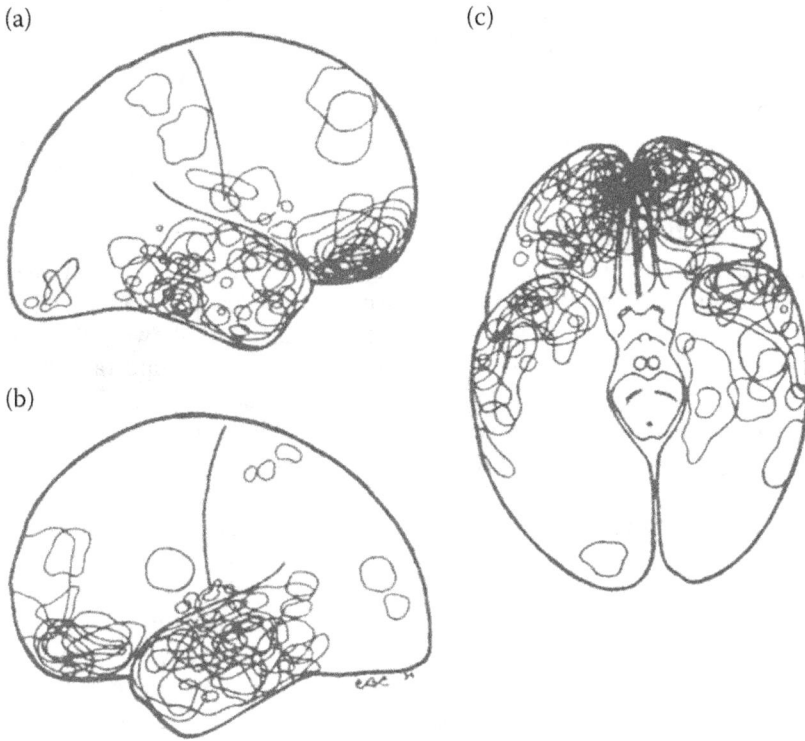

Figure 5.1 Courville (1937 and 1950) line drawings of brain damage.

In this regard, if the brain is slapped hard enough against the front of the skull in linear motion, the classical coup-contrecoup traumatic brain injury occurs. Thus, in large effect the main force is transmitted initially to the frontal and temporal lobe structures (coup), as the brain "bounces back" atop the stalk of the spinal column it is slapped against the back of the skull (contrecoup) and potentially occipital lobe damage occurs. This depends on the speed of the acceleration and deceleration, for as I have stated, the skull does not have to be struck, it is the speed, and then the speed of decline in the speed, that are the critical factors here. Potentially, apart from Courville, the most important historical work in this regard is that of Adams, Graham, and Gennarelli (e.g. 1983), it is hard to think of a better explanation, they explicitly stated:

> All of the principal types of brain damage that occur in man as a result of a non missile head injury, viz. cerebral contusions, intracranial haematoma, raised intracranial pressure, diffuse axonal injury, diffuse hypoxic

damage, and diffuse swelling have been produced in subhuman primates subjected to inertial, i.e. non impact, controlled angular acceleration of the head through 60° in the sagittal, oblique and lateral planes.

Functionally, this will lead to well-understood consequences, as will be shown later. For now, the essential point is the difference between focal and diffuse injury in TBI is easy to state but harder to make a useful clinical rationale for doing so. Indeed, diffuse axonal injury is the *sine qua non* of TBI. Thus, although focal injury may be apparent (e.g. specific site of subarachnoid bleeding) and the most frequent immediate contusion sites have been highlighted, the larger effects potentially are how and in what ways the axons themselves have been affected. As has been shown follow-up studies in children and adolescents have demonstrated important later atrophy in the corpus callosum (CC) due to the force-displacement (downstream) from the initial coup injury. It is almost like a shock wave from a bomb blast or a stone flung in water, damage travels along the axons. This atrophy is transmitted damage through the axons connecting the frontal and temporal lobes to the CC. The diffuse damage does not finish at that junction. If one thinks again about the position of the brain atop the "pole" of the spine, in conjunction with the projections of axonal connections; top to bottom, side to side, back to front, and the reverse of each of these planes, diffuse axonal injury is bound to be one of the central kinds of primary injury in TBI. This of course potentially disrupts any number of minor roads, major roads, up to super motorways of axonal connections involving important systems, as opposed to single lobes. To take one example, in a recent paper by Owens et al. (2018), using fractional anisotropy (a specific DTI) they found significant associations between the degree of white matter damage in the: ' ... corpus callosum, superior longitudinal fasciculus, cingulum, inferior fronto-occipital fasciculi, corona radiata and cerebral white matter,' and attentional problems, but most clearly through slowed information processing speed post-TBI.

The complicated nature of TBI is made further problematic through the possibility of a skull fracture or fractures. Whilst this may not inevitably lead to a more severe injury or consequence, unless it is treated carefully, the two most common negative outcomes are post-traumatic seizures and or brain infections. The latter is even more likely if the fracture is a depressed one. In this case, tiny bone splinters may pierce the tough dural layer, and if large enough may even pierce brain tissue, in each instance they bring the potential for a bacterial entry point. Lin et al. (2015), argue that the actual incidence of subsequent bacterial infections is frequently overestimated. However, they go on to state that when an infection of this kind does occur, it almost inevitably leads to a much poorer neurologic prognosis, final level of morbidity, especially cognitive issues, and mortality is higher. Indeed, quoting the oft-cited study by Harrison-Felix et al. (2006), they conclude that infection-related mortality rates are around 28%. Although classically fractures in the skull are associated with greater risk of bacterial infection after a TBI, this was not found by Lin and colleagues, instead the greatest risk appeared to be from CSF leakage, even more prominently drainage of lumbar and ventricular areas were independent risk factors, outweighing even multiple craniotomies (Lin et al., 2015). Having said this most authorities would agree that some form of penetration of the dura, will increase the risk of bacterial infection, yet there is little actual evidence for this. Indeed, recent studies suggest a greater risk of infections through CSF leakage or drainage, as with Lin et al. (2015). For example, Prakash et al. (2018), concluded there were no significant associations with meningeal tearing and furthermore, no significant data to support any prophylactic antibiotic usage.

This is crucial as any use of antibiotics should be restricted, as it takes so long for our biome to recover. Moreover, as long ago as 1972, when Jennett and Miller found only a 10% infection rate amongst a large group of civilians with compound skull fractures. They asserted that incomplete debridement was the primary reasons for this and that antibiotics are: ' ... no substitute.'

The conclusions would appear to be that it is unquestionably the case that infections produce worse outcomes after TBI, but that these are more likely to be the result of other less obvious factors than fractures *per se*, indeed the most likely reason would be for a prolonged CSF lumbar drain. However, the argument regarding the biome and its need for protection is crucial.

The same cannot be said for epilepsy and TBI, where robust data exist for increased risk of developing epilepsy is apparent. Risk for early and late development of seizures is increased further for those with intracranial metal fragment retention and depressed skull fractures (amongst other risk factors). This has been summarized recently in translational research and clinical practice in traumatic brain injury, by Ding, Gupta and Diaz-Arrastia (2016).

What appears to be the case is the counterintuitive idea for the layperson and the trainee clinician; that as the injury "develops" the consequences become more debilitating. In that, contusions or bruising cause problems, the deeper problem resulting from the initial physical force causes damage to axons across the whole of the brain parenchyma, as it is not simply the axons and thus pathways and tracts that are damaged but also these pathways endpoints and stopping points along each pathway. Potentially deeper and even more widespread damage occurs through the disruption to the neurochemistry previously discussed (cf: new damage, free radical, neurotransmitters, neuromodulators, and neuroinflammatory processes). Brain injury, like sleep, is a complex process. Indeed, it is a complex of multiple complex processes. As importantly the implications of the foregoing strongly attest to the fact that a single "magic bullet" pill or overly simplistic solution to either is not just unhelpful, it is potentially dangerous to think in such terms.

Does this mean that the damage associated with hypoxic or stroke damage is any more understandable or at least less wide-ranging? The short answer is most definitely yes, the long answer is yes, but not quite.

Stroke

There are clear localization commonalities in stroke, even though it is generally associated with a diverse range of clinical manifestations. Thus it is generally held that lacunar (from lacunae: crescent or moon-shaped) infarcts are most frequently found in subcortical regions. And they account for something like 20% of all strokes. They are likely to be multiple and small (1 to 20 mm) in size; frequently found in the arteries supplying the deep white matter areas, the pons, internal capsule up through the basal ganglia and the thalamus. Most commonly these kind of strokes are associated with undiagnosed hypertension and have become synonymous with cerebral small vessel disease. So numerous in their appearance on MRI they have become associated with subcortical and brain stem structures that can be seen in early cases of preclinical dementia. If we recall the discussion about the importance of white matter changes and diffuse axonal injury in TBI, recent work on lacunar infarcts even in deeper areas of the brain makes much better sense. By this, it should not mean the old (or indeed modern!) view of diaschisis holds true here. For example, Hoffmann and Schmitt (2004) looked specifically at brain stem lacunar infarcts and associated white matter changes for which they found evidence of distinct executive impairments. At that point it should be noted that this has implications for disconnection syndromes between the axonal connections to and from the brain stem and frontal lobes; rather more akin to the earlier discussion in TBI around disconnection syndromes. Having said this it is far more likely that lacunar infarcts will produce pure motor hemiparetic or sensory-based deficits. This is primarily a consequence of the neuroanatomical location being areas such as the brain stem, pons, internal capsule, and thalamus. Within this conceptualization, it is also pertinent to draw the contrast between clinical impressions and research data; as it is a commonplace that lacunar infarcts are the inevitable precursor to vascular dementia. They should be viewed, in a more nuanced way, more of an increased probability, a bellwether. They are most likely an indication of small vessel disease, which is then even more likely going to lead to vascular dementias. The conclusion from Makin et al. (2013), and other more recent papers (for example: Sexton et al., (2019) is that the actual figures are more likely to be between and 24% and 39%, which whilst extremely high, are not the inevitable conclusion we may have thought.

Unlike other forms of brain injury, strokes are now more commonly referred to in terms of the main arterial territory affected, or cause. The starting point for understanding these ideas is the underlying division between the cerebral hemispheres, both of which are supplied by the carotid and anterior circulatory systems and the brain stem and posterior aspects of the hemispheres, supplied by the posterior and vertebral basilar circulation. Even here this helpful distinction breaks down upon further inspection, as with most rules of thumb, as aspects of the posterior hemispheres are supplied by some more anterior circulatory systems and equally posterior systems can supply more anterior or at least medial aspects of the hemispheres. In any event one must understand the circle of Willis, as the key area which supplies all the major arterial systems to the brain. This beautiful illustration from Gray's anatomy demonstrates large middle cerebral artery in comparison to others (Figure 5.2).

The doyen of neuroanatomy and clinical disorders, Adams et al. (1993), suggests that carotid and anterior circulatory strokes commonly manifest clinically as sensory loss and visual field cuts; especially homonymous hemianopia. To illustrate these considerations further, a brief look at middle

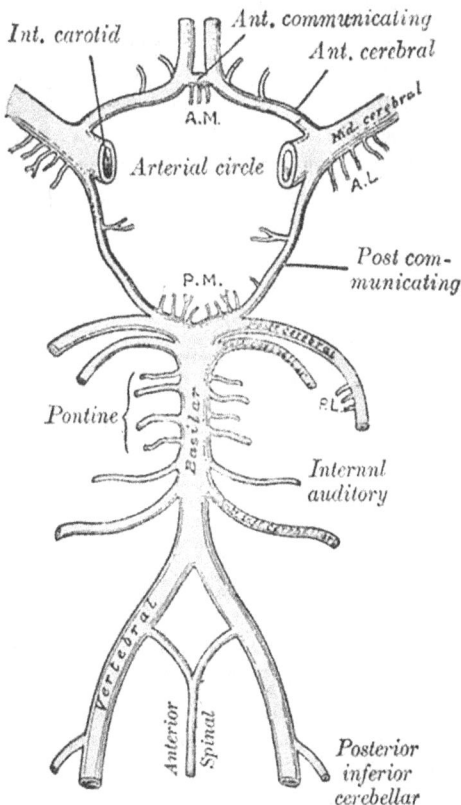

Figure 5.2 The Circle of Willis from: Henry Gray (1918) *Anatomy of the Human Body.*

cerebral artery (MCA) strokes, will be made. The MCA's cover such a large aspect of each cerebral hemisphere strokes can here cause a wide range of clinical presentations. The territory under discussion covers large aspects of the medial frontal lobes, parietal lobes, internal capsule, and the majority of the anterior CC and the head of the caudate. In other words, something like 70% of each hemisphere is supplied by this arterial system. Understandably, as it supplies areas of the prefrontal and parietal lobes involved in sensory and motor processing, it can manifest as contralateral upper and lower limb weakness and sensory changes. The diagram below demonstrates the size of the territory the MCA covers.

Frank Gaillard. Patrick J. Lynch, Medical illustrator (Brain_stem_ normal_human.svg) CC BY-SA 3.0

In Navarro-Orozco and Sanchez-Manzo's (2019) beautiful paper entirely devoted to the middle cerebral artery, they detail its key structures and functions. I cannot better their detailed overview, hence:

> The cortical branches of the MCA irrigate the brain parenchyma of the primary motor and somatosensory cortical areas of the face, trunk and upper limbs, apart from the insular and auditory cortex. The small central branches give rise to the lenticulostriate vessels, which irrigate the basal ganglia and internal capsule. The superior division irrigates the lateral inferior frontal lobe, which involves the Broca area responsible for speech production, language comprehension, and writing. The inferior division of the MCA irrigates the superior temporal gyrus, which involves Wernicke's area responsible for speech comprehension and language development.

They go on to describe some of the more common clinical manifestations of vessel problems. To begin, as with Adams et al. (1993), they describe MCA arterial strokes caused by emboli. Typically, these arterial occlusions are caused by a dislodged thrombus from the heart. These are thus seen commonly after a myocardial infarction (up to four weeks later), mitral valve stenosis, prosthetic valve replacements, and other heart diseases. However, very frequently these are not caused through the usual process of atherosclerosis, but by atrial fibrillation. The extreme fluttering causes a temporary thickening of the blood and the formation of a thrombus or emboli. In addition, internal and common carotid atherosclerotic plaques may break off and travel up through to the MCA. Upon entering the brain these lodge in the MCA; if the thrombus or embolus is large enough to get stuck in the wider parts of the system, i.e. it is proximal, it can cause enormous damage, as the various upstream smaller vessels will become starved of oxygenated blood and thus cause a huge potential penumbral area. If distal, and potentially through a smaller diameter thrombus or emboli, the resultant ischaemia and final necrosis will be smaller as it will have travelled further into the smaller vessels of the system.

This particular type of ischaemic stroke produces what many understand to be the classical or stereotypical presentation. Indeed, its name is a middle cerebral artery syndrome and presents with contralateral sensory problems because of tissue damage to the somatosensory cortex. Palsy of the upper, and lower limbs, and the face is also commonly seen due to the motor cortex. Clinically paresis, spasticity, hyperreflexia, and upper motor neuron signs are also common. Some clear dominant and non-dominant differences can be found, including: significant speech and language issues stemming from a Broca or Wernicke type aphasia in the dominant hemisphere. In the non-dominant hemisphere apraxia, anosognosia, and hemispatial inattention are frequently found on clinical examination.

These typical presentations may be differentiated with the less common haemorrhagic Charcot-Bouchard micro aneurysmal presentation. No less clear on the scan they appear to be the result of chronic hypertension. The risk factors for this are the usual combination of modifiable and non-modifiable lifestyle choices. In that high-fat diets, high waist to hip ratio, smoking, high alcohol consumption, poorly controlled cholesterol, or cocaine (amongst other) drug use. Unalterable factors include race and ethnicity (greater incidence in African and Asian populations), being male and increasing age. Thus, the risks are similar but the presentation is different and increasingly the case. These type of ever-expanding (c.f. The Law of Laplace) mini aneurysms eventually burst, primarily in the pons, cerebellum, thalamus, putamen, and caudate areas.

Aneurysmal stroke, the most common type, of the larger kind, is known as berry (saccular) present in branch bifurcation or trifurcation points. They usually begin as a small weakness in the vessel wall which then becomes a small but expanding pouch. About a third of these type of aneurysms are found in the MCA. The neurosurgical treatment of these through coiling or previously clipping is considered by one neurosurgeon, as one of the very few neurosurgical treatments of genuine advance since the 19th century (Marsh, 2014).

Summaries of clinical presentations can be found in standard texts such as Kuybu et al. (2020). Another clinical feature is that although anterior strokes are far more common, around 20% of strokes occur in the posterior cerebral arteries (PCA). In addition, the three most frequent causes are atherosclerosis (48% if large and small vessel disease are included), embolism, and dissection. The latter, although the rarest form in clinical practice, can be the result of chiropractic manipulations and higher-risk jobs, such as plumbers, kitchen fitters, and aircraft technicians, as well as the more usual trauma from a vehicle accident (Nouh et al. 2014).

Hypoxia

Unsurprisingly, hypoxic brain injuries follow the arterial system. However, they are less understandable from that perspective, due to variations in patency associated with cumulative modifiable lifestyle choices.

Most commonly hypoxic injury is the result of cardiac arrest, vascular injury, strangulation, failed hanging, intoxication of smoke or other toxic gases, drowning, and drug overdoses. Undoubtedly, the most frequent clinical presentation is after a cardiac arrest. As described earlier the brain is a very greedy organ, it consumes constant high amounts of oxygen and glucose, as it is unable to store any form of energy, and any variation in this supply will have deleterious effects rapidly. This is especially true after the brain switches to poorer anaerobic production methods, as they are not only insufficient to meet demand, they also produce far more waste products, exacerbating damage. People think that as commonly viewed "good fats are good for the brain", it can switch to these for energy. Unfortunately, in terms of flexibility, this does not actually work. The brain can utilize the ketones produced from fats, but these nutrients are not a whole solution for the majority. A strict ketogenic diet does however, work well for epilepsy and was, prior to the convenience of taking an anti-epileptic pill, the treatment of choice in the early part of the 20th century (see for example: Wheless, 2008). It is thought that so-called good fats (e.g. from fish) do not necessarily have direct effects on the brain, it is rather their indirect effects. Some of these were discussed well by Gomez-Pinilla (2008) in a summary paper. He suggested that Omega-3 fats, and specifically a sub-component, known as DHA is an essential part of neuronal cell membranes, but it cannot be internally produced; so has to be ingested from our diet. In rat models, increasing DHA supplements, elevate levels of hippocampal brain-derived neurotrophic factor (BDNF), which is an essential element of brain plasticity, which enhances synaptic membrane fluidity, which in turn enhances cognitive abilities. Other than the modifiable lifestyle choices producing oddities in presentation, for example, smoking damage to particular arterial structures and not others, the neuroanatomical structures damaged are largely similar. Most frequently the so-called watershed areas are harmed. These areas are generally held between the anterior and middle cerebral artery regions. The point here is that blood supply is decreased, due to the areas being distal from larger arteries. They are thus vulnerable as blood supply is decreased, so they are predisposed to relatively small changes in nutrient density and hypoperfusion. They are often subdivided between cortical (CWS) and internal watershed (IWS) areas. CWS regions include the fronto-parasagittal regions from the frontal to the lateral ventricle. They can also affect a strip along the Sylvian fissure. Conversely, and less commonly they can develop from the lateral ventricle to the parieto-occipital cortex as a posterior CWS. Classically, IWS is seen as a rosary-like chain around the centrum semiovale (CSO). Larger IWS come alongside the lateral ventricle. Whilst this explanation may seem straightforward it must be noted there is wide inter and intra-individual variation. What is clearer from perfusion studies using sophisticated PET, Xenon computerised tomography (CT), and single-photon emission computerised tomography (SPECT) protocols is that the heart of haemodynamic abnormalities were in evidence for anterior CWS but not posterior CWS. Moreover, CSO rosary pattern have been shown to be highly associated with haemodynamic impairments (see for example: Momjian-Mayor and Baron, 2005). Indeed, as long ago as Krapf et al. (1998) these patterns were proposed as a consequence of brief declines in blood pressure and other patterns as potentially associated with more chronic hypoperfusion. What is even clearer is that the areas described and the mechanisms underlying them are even more obvious and pronounced in hypoxic brain injuries. Scans demonstrate these findings depending on the time period from the hypoxic episode. Initially, swelling is demonstrated, followed by hyper signal areas in the basal ganglia, white matter degeneration, cortical laminar necrosis, and finally atrophy. These changes follow the supposed four phases of hypoxic damage from hours to days then weeks.

It is for these reasons that primarily temporal and frontal structures are damaged cortically. Sub-cortically, the basal ganglia are frequently damaged. Indeed, pyramidal neurones of the hippocampus, cerebral cortex, and lower in the caudate and putamen as well as the Purkinje cells in the cerebellum are affected. Pyramidal neurons are heavily weighted towards excitatory neurotransmitter receptors. One of the hallmarks of hypoxic presentations is the triad of frontal, memory, and attentional problems alongside a very easy-going personality. Indeed, the latter can present as almost a complete happy-go-lucky, giggling attitude to life, with much of the drive having left the individual concerned. Unfortunately, this would largely be considered a good outcome. Most studies reveal what has been described as a 'dismal' outcome in regards to morbidity, and with a high 64% mortality rate, the very firm conclusion was that the best thing is to prevent this type of brain injury from happening in the first place (Lacerte and Mesfin, 2019). This leads on to some of the functional characteristics seen in some of these brain injuries.

Chapter 6

Functional disturbances caused by brain injury

This chapter will consider the functional results of the brain injuries discussed in the foregoing chapters. These will include but are not limited to physical symptoms such as: headaches, difficulties with balance, visuoperceptual problems, seizures, problems in understanding or expressing language, changes in sexual function, and motor impairments.

These are most commonly associated with other debilitating and potentially less easy to adjust to problems such as: personality changes, difficulty with reason, planning and logic, memory impairments, depression and/or anxiety, poor attention, disorientation, and acting inappropriately.

It is only more recently that awareness of the full gamut of long-term consequences of acquired brain injury (e.g. even in milder forms of TBI) has begun to be recognized. As previously discussed the damage to the neurochemical systems and axonal connections running from rostral to caudal, especially the circuitry in and through the hypothalamus will produce profound shifts in thinking.

Thoughts on neuroinflammation

First, acquired brain injury of all kinds is associated with increased morbidity and mortality. According to American data produced by the CDC 22% of survivors of a TBI die within five (5) years of the index event. Those who suffered a moderate to severe TBI were likely to have an overall shorter lifespan than those who hadn't experienced such injuries; on average by nine (9) years (CDC, 2017). Even those with mild TBI have been found to have damage to the CC some three months after such an injury; which clearly points to much greater potential for increased morbidity than was previously acknowledged, indeed if oedema is added to the algorithm they were also five times more likely to die than those who didn't demonstrate any oedema during early hospitalization (Tucker et al., 2017).

As the recovery hopefully gains pace, if the individual progresses, further challenges may become apparent. Once again some of these outcomes appear to be related to severity but others are much more related to the host

DOI: 10.4324/9780429199066-6

response and the individual's inflammatory reactivity. A number of these have been specifically studied, interleukin-6 (IL-6) is one such. In 2015, Kumar and his colleagues in Pittsburgh studied this pro-inflammatory cytokine in relation to functional outcomes at 6 and 12 months. They found that remarkable differences were apparent depending on the initial total inflammatory load. In other words, severity effected functional outcome, but an additional factor was found to be in play if there was also poly-trauma. The same severity of TBI had different functional outcomes if poly-trauma (other significant areas of bodily or physiological system damage, in addition to the TBI) was also apparent. These latter cases had significantly raised CSF levels of IL-6 during the acute sampling phase of the study. In an earlier study, Jeungst et al. (2014) studied a different inflammatory marker and its specific relationship with disinhibition and suicidality. Tumour necrosis factor (TNF-α) was sampled (CSF and serum sampled) in moderately to severely injured TBI patients from the time of their injury through to some 12 months post-injury. Two standardized measures were utilized at six months and 12 months post-injury. These were Frontal Systems Behaviour Scale – Disinhibition Subscale and the Patient Health Questionnaire (PHQ). Acute and chronic TNF-α was significantly associated with disinhibition at six months post-injury, and six-month disinhibition was associated with suicidal endorsement at both six and 12 months and disinhibition at 12 months post-injury. The problem with this study is the usual correlation problem of not necessarily being to do with causation. The other problem is the very structures implicated in disinhibition and suicidality are the frontal areas and connected pathways associated with moderate to severe TBI. There did seem to be a largely independent effect irrespective of severity, but it was less clear than Kumar et al.'s (2015) study. Having said this Postolache et al. (2020) cite overwhelming evidence from decades of research on animals and more recent research in humans, that points to various inflammatory processes that go awry in acquired brain injury (especially TBI). They go on to describe studies similar to Kumar et al.'s (2015) and Juengst, et al.'s (2014) that contribute specific deficit understanding with specific inflammatory processes. One example of such specific functional outcome work is that of Bodnar, Morganti and Bachstetter, (2018). In this review, they aim to better understand the link with inflammatory processes and depression. Citing the decades of research on inflammatory processes in depressed patients without brain injuries they then turn to two specific markers that may help explain this phenomenon better in the brain-injured population. These were TNF-α and interleukin-1. They assert that these pro-inflammatory cytokines peak during the acute stages of TBI recovery, but persist for months and years after this, during which time post-injury depression is also at its peak. They further propose that the most likely mechanism is direct signal alterations at the neuronal synapse. The evidence, they suggest: ' … although largely associative is compelling.'

Similar findings have been found for relationships with inflammatory processes after TBI and the extent of cognitive dysfunction. Given the earlier discussion, what is of most interest in this emerging literature is, once again, the apparent non-linear relationship between either severity of the initial injury or the site of initial impact and functional outcomes. Instead, the importance of inflammatory processes and the chronic microglial activation previously examined (through a priming effect) may be a more powerful way of addressing the functional outcome conundrum. To this end, Ramlackhansingh et al. in (2011) demonstrated these ideas clearly. In a small but well-designed study, ten patients were followed through assessments of PK binding (an indicator of microglial inflammatory processes). The most severe cognitive impairments were found in those patients who had the most sub-cortical damage; especially in the thalamus, irrespective of initial injury site. The other main finding was the length of the chronicity, as they concluded:

> We demonstrate that increased microglial activation can be present up to 17 years after TBI. This suggests that TBI triggers a chronic inflammatory response particularly in subcortical regions. This highlights the importance of considering the response to TBI as evolving over time and suggests interventions may be beneficial for longer intervals after trauma than previously assumed.

In a recent review, Javidi and Magnus (2019) argue that alongside the previously discussed protective and regenerative effects of inflammation there is increasing evidence of a detrimental maladaptive immune responses. Most specifically post-stroke measurement of B cell and T cell clusters correlate with cognitive decline. This builds upon numerous other investigations which amply demonstrate significant correlations with the degree of autoimmune response, the subsequent size of the lesion, and final functional outcome in stroke and TBI. This paper also reviews the effects of specific types of immune and autoimmune phases at different time points after the brain injury. It highlights why earlier research on simply trying to reduce the overall inflammatory responses with broad-spectrum drugs (e.g. prednisolone) have had such mixed results. Fine-tuning how the host modifies its own response is a far more helpful approach. One such avenue of research comes from the more recent understanding of regulatory T cells (Tregs). One of the functions of Tregs cells appears to be controlling autoreactions; specifically they seem to inhibit some autoimmune functions, maintaining greater tolerance for disruption before allowing full autoimmune reactions to occur (Dominguez-Villar and Hafler, 2018). The largest review to date confirmed the majority of research has demonstrated this supportive role. Indeed, Tregs depletion studies consistently demonstrate larger infarcts, increased leucocytes, and far more pro-inflammatory cytokines. The converse, substantial amounts of Tregs, have shown better overall neuro-logical recovery in stroke patients (Ito et al., 2019). Similarly, Johnson et al. (2013), found in almost a third of cases studied had persistent microglial activation well over a year post-injury from a single trauma. They go on to argue that inflammation and concomitant white matter degeneration persist for years. Especially noteworthy is the CC thinning they found, which has obvious parallels with the paediatric cases noted earlier. To this end Henry et al. (2020) demonstrated through experimentally depleting microglial ac-tivation during the chronic phase of TBI reduces neurodegeneration and neurological deficits. Using adult mice they "switched off" the microglial activation and thus neuro-inflammatory loop at one month post controlled cortical impact and then "switched it back on" later. Thus, an absent period, which they proposed was the critical time period that microglial activation occurred was halted. Various measures were then found to result in potential reductions in neurodegenerative processes that are frequently accompanied by neuroinflammation. Furthermore, they argue strongly that the ther-apeutic window for TBI is potentially far longer than traditionally thought. This was especially true if 'evolving' later neuroinflammation were to be wholly inhibited or at least, better controlled.

Finally, and most pertinently for this chapter they found that the treated mice had far better long-term motor and cognitive functions. Moreover, before some hue and cry occurs regarding mouse and rat models, unlike for pharmacodynamics and pharmacokinetics, most of these have stood the test of time and human response. Indeed, one only needs to think of Adams, the doyen of TBI research on white matter damage, who made these points robustly in animal research well before they were confirmed in later human studies.

Therefore, host idiosyncrasies and immune responsiveness add to this complexity. What is clear is that the total immune response can become magnified (e.g. poly-trauma) leading to greater a pro-inflammatory cytokine storm and bigger autoimmune effects which have direct effects on lesions and final functional outcomes.

Bones

As has been discussed, depression, cognitive, and motor problems are abundantly researched and discussed. This book is not entirely about functional outcomes, there are numerous ones the interested reader could consult. As ever, in this regard, I am keen to use selective functional out-comes, rather than an exhaustive list of potentially shallower but more wide-ranging functional outcomes. A less well-known long-term functional out-come of TBI is that of bone density and other skeletal problems. The rea-sons behind this are complex but in large measure encompass the hormonal system and the neural circuits originating in the hypothalamus. As such these long-term effects can just as easily be very distal to the site of injury. Bear with me as I will explain later why this links to what has gone before.

One of the ways in which these systems can go awry has been examined by Sullivan et al. (2013), in their review. Neurogenic heterotopic ossification (NHO), is one of the three ways (the others being genetic and traumatic) this problem can arise. NHO has been found to be quite common in the brain-injured population. It has been estimated to effect over 20% of those who suffer an acquired brain injury or spinal cord injury. In essence, it is the formation of bone structures around non-bone structures (muscles and soft tissue). It is most common around the musculature of the hip, elbow, knee, and shoulder. It is excruciatingly painful at a certain point in growth. It has been suggested as a tissue repair process that has gone wrong. It used to be thought, in rather simplistic terms, as a problem of a lack of mobilizing (ne: "effort"), with all the attendant blame of the patient or the physiotherapist involved. However, recent studies have accumulated a deeper understanding. One such was the recognition that something unusual happens to indirect signalling after a traumatic brain injury. In this regard, serum was collected from muscles after a TBI and was compared to non-TBI controls, in addition to simple bone fracture patients. Cultures were taken from the same muscle groups in each of these patient populations. There was a clear and significant difference between each group, with the controls having few markers, the fracture group having some, and the TBI group having a significant increase in osteoblast markers. Indeed, the study by Cadosch et al. (2010), concluded that:

> Human serum supports the osteoblastic differentiation of cells derived from human skeletal muscle, and serum from patients with severe traumatic brain injury accelerates proliferation of these cells. These findings suggest the early presence of humoral factors following traumatic brain injury that stimulate the expansion of mesenchymal cells and osteoprogenitors within skeletal muscle.

The exact mechanisms by which this increase in osteoblast expression occurs in muscles and soft tissue remains elusive, but one hypothesis is it is another result of disruption in the electron transport chain, which was unpacked earlier. In effect, the hypoxic conditions that occur after most acquired brain injuries at the neurochemical level led to many functional deficits, one of which can ultimately be heterotopic ossification. The specific form this takes is the angiogenesis goes into a kind of inappropriate overdrive to compensation for the hypoxic reactions (and thus increases in reactive oxygen species) leading to hypertrophy, this leads to altered signalling which then makes muscle and soft tissues increase (through this and other humoral factors) osteoblast formation.

Similar patterns of over-stimulation and resultant inappropriate enhanced osteogenesis can occur after any kind of post-TBI bone fracture. Or as Locher et al. (2015) describe the process as: ' … exuberant callus formation.' In their confirmation of clinical data study on mice they found the TBI group had: ' … increased bone volume, higher mineral density, and a higher rate of gap bridging compared to the fracture group.'

It seems counter-intuitive to think that brain and bone are linked, but from another view, how can it be otherwise. There is now a deeper knowledge of the bidirectional relationship between the two. Trauma can be viewed as effecting both profoundly, and most clearly not simply through direct damage to either. This can be seen in a number of neurodegenerative diseases (e.g. Alzheimer's) but also acquired brain injuries such as TBI and stroke. Post-stroke fractures are common. The reasons used to be thought of as a general fragility, and was deemed (yet again, there is a theme here) a problem of poor mobilization, poor physiotherapy, or poor motivation, or all three! However, now it is recognized as more likely to be as a result of disturbances in Vitamin D and related calcium depletion, in addition to hyper osteoclast activation. The latter and other markers found (e.g. elevated sclerostin, osteopontin) which potentially mean direct stroke effects and stroke endocrine effects. These indicators lead to further thoughts around melatonin and its levels after brain injuries, in addition, to the more obvious hypoxic conditions following these injuries previously discussed, leading to heterotopic ossification and neuroinflammation exacerbating secondary damage.

Other than the widely discussed hypoxic issues I've charted, another more familiar mechanism has been suggested. Melatonin becomes crucial to understanding bone health. Melatonin regulates calcium balance and bone metabolism. It promotes the proliferation of osteoblasts and stimulates bone mineralization, amongst many other activities. It is also plentiful in bone marrow, some think acting as an important precursor of bone cells. Not only this, but it would appear it plays a significant role in osteoclast (bone dissolving and reabsorption) activity.Moreover, melatonin production is frequently disrupted in brain injury, leading to a wide range of melatonin-related disorders. In traumatic subarachnoid haemorrhage there is a known increase in melatonin during the initial phases of recovery, which some have hypothesized leads to the fast healing seen in fractures, and heterotopic ossification in this sub-sample (see for example, Alu, Kiselev and Loboda, 1977; and Huang et al., 2018). The role of melatonin in bone health will be discussed further in regards to treatments. Suffice to say it (either excess or deficit) is certainly implicated in functional outcomes in those with a brain injury.

Memory

More frequently understood functional consequences of acquired brain injury are various forms of memory problems. I will look at this using TBI as an exemplar. The reasons for this are the intimate linkage between these functions and sleep, which have previously been explored. The intention is also not to be thought of as a definitive account and examination of the state of play in memory research and theory, which is far beyond the scope of this chapter. In any event, there are many superb texts designed for that purpose (e.g. Lieberman: Learning and Memory, 2nd Ed, 2020; Gluck, Mercado & Myers, Learning and Memory: from brain to behaviour, 4th Ed, 2020; Baddeley, Eysenck, and Anderson; Memory, 3rd Ed, 2020). The aim is rather to sensitize the reader to important aspects of the topic of attention, learning, and memory that can cross-fertilize the previous work on memory and sleep and the later parts of damage to the brain and its potential for amelioration.

I well remember listening to the great South African neuroscientist Mark Soames tell an aghast audience at a conference a few years ago that really all mental functions (including attention, executive, and everything else) are really different forms of memory. Needless to say much debate and argument ensued. This is pertinent because whatever we may feel and think about that statement, memory provides the great binding force of our mental life, without its proper functioning we cannot have a sense of self in our past, or indeed as profoundly, in our potential future. It may be the current therapy *de jour*, but being completely mindful, in the moment, is actually such a burden of contextless emptiness. Our most damaged patients become a shifting photographic machine without any accompanying album or context to hold this "empty" picture in time. The fragmentary nature of a truly destroyed memory system(s) can be horrifyingly seen in the classic cases of the patients: H.M. and Clive Wearing (Scoville and Milner, 1957 and Sacks, 2007). However, in the usual rough and tumble of clinical casework, less dramatically perhaps, we are weekly witnesses to the anxiety-provoking menaces that profoundly damaged memory systems wrought.

Although memory and its related concepts of learning and attention have been pondered and written about for millennia, it is widely acknowledged that it was only since scientists examined the sad and unusual case of H.M. in the middle of the 20th century, that it became better understood, most especially parts of the neuroanatomical substrates of memory. Since this period it is fair to say that the last 70 years have produced an explosion of research, theory, and clinical material regarding memory. The critical early period suggested the hippocampi in the medial temporal lobes, completely absent through neurosurgery for H.M., were profoundly important for proper memory functioning. From these nascent neuroanatomical understandings, parallel theorizing produced early cognitive models of memory. In essence, these could be viewed as unilateral and highly neuroanatomically discrete. Memory was viewed as a floating ordered system largely (solely?) involving the hippocampi and often linear in conception. The best and most long-standing models of these early models were produced by Atkinson and

Shiffrin (1968) which conceptualized three structural components. In order these were the sensory store, the short-term memory store and then the long-term store. From this developed the levels of processing model (Craik and Lockhart, 1972) which suggested memory works from shallow to deep processing; the latter being more robustly remembered. This became subsumed in the earlier model. After this, Tulving's models (1972 onwards) based on the kinds of materials that are stored developed. In this regard, episodic memory-related to experiences and their relationship with events in the world; semantic memory related to the facts or knowledge of the world, and finally procedural memory involves knowing how to do something, it is skill-based (riding a bicycle, making a cup of tea). Again, these kinds of definitions have now been subsumed in more complex models. However, as descriptors they are still much in evidence, but not as models in and of themselves. It is widely held that declarative memory is more readily seen as a hippocampal and peri-hippocampal cortical event or largely seen as such. This would also be true for further fractionated understanding, based around Baddeley's various formulations of the specifics of STM, as working memory (e.g. Baddeley, Eysenck and Anderson, 2020). The so-called visual part of the visuo-spatial scratch pad, and the auditory-verbal one being designated the phonological loop, and its overarching manager a mini executive system. Whereas, implicit (unconscious) memory is viewed as taking place in other systems, such as the striatum, the cerebellum, and the amygdala. Finally, as with our greater understanding historically, and previously discussed, of the neuroanatomical substrates of external and internal behaviours, theories and models of memory have evolved into rich, complex systems and distributed or parallel models.

Indeed, Eichenbaum and Cohen (2004) argue that all of the previous discussions on memory and the previously described models are either wholly incorrect or at best overly simplistic. Furthermore, they strongly argue that: 'Memory is a fundamental property of brain, and its storage is intimately tied to ongoing information processing in the brain; and Memory is manifested in multiple ways by multiple, functionally, and anatomically distinct brain systems.' The key distinction in this much more complex understanding of memory is between: '...conscious and unconscious forms of memory and that each of these are wholly derived from the neuroanatomy and neurophysiology involved in each system.'

Those more familiar with the work of Thorndike, Hebb, Tolman, and Bartlett will more readily see how more recent research has come to this point, of flexible re-constructive, fluid "memory making" and complex internal system approaches rather than the linear flowchart approaches beloved of the early cognitivists.

Within such frames of reference, it may appear that the hippocampal centrality has been lost. Whilst it is true that much more recent focus has been placed on non-hippocampal memory systems (e.g. mechanisms of relational memory) knowledge of the hippocampi has actually deepened. Indeed, in one simple way research has confirmed that the dentate gyrus in the hippocampi is one of the most fecund brain areas for stem cells. The other areas that are especially important for memory and other essential cognitive functions are: the hypothalamus and the underneath part of each lateral ventricle. Very few stem cells are needed for considerable new neuronal development, but if these areas are damaged then the potential for plasticity is clearly going to be diminished (Andreotti et al., 2019). Thus the self-same areas needed for adequate memory functioning are those also supremely important for neuroregeneration and thus recovery. They are also very likely to be damaged in the faulty biochemical cascade described earlier and in later chronic inflammatory processes.

Much of the earlier knowledge came from bottom-up approaches to understanding learning in other species. The time honoured behavioural concepts of classical conditioning and related areas promoted by Pavlov and Thorndike; found experimental evidence in invertebrate studies of fruit flies and the like (e.g. Menda et al., 2011). In each form of simple organism learning there was evidence of concomitant short and long-term memory. The conditioned stimulus didn't simply temporally occur prior to the unconditioned stimulus, through basic memory mechanisms it eventually predicted it, making it unsurprising to the organism (e.g. Carew and Sahley, 1986). From here studies attempted to understand the synaptic connections between sensory and motor neurons. Sure enough the conceptual framework of stimulus releasing serotonin on the sensory neuron, leading to increases in cyclic adenosine monophosphate (cAMP) which signals the release of glutamate into the synaptic cleft. The result is a temporarily strengthening of the sensory-motor connection; eventually if repeated, leading from short term to longer or more permanent strength connections. The sequences are largely similar and have been measured for short, medium, and longer-term memory synaptic strengthening procedures (e.g. Mayford, Siegelbaum and Kandel, 2012). This process is known as memory consolidation. To summarize key neurotransmitters in this process are the right amount of serotonin and glutamate, with yet another adenosine-dependent process linking them. This phase involves protein changes which result at the cellular level in different waves of gene expression and novel neuronal plasticity which coalesce into new synaptic connections and eventually some level of permanence in memory systems (Puthanveettil and Kandel, 2011). Neurones can have many synapses (a few to perhaps a thousand in some cases) and it is at the level of the synapse that a unit of short, medium, or long-term memory occurs. It is only during the course of the early 21st century that we have been better able to understand these essentially epigenetic mechanisms in supposedly immutable adult mammalian brains (Levenson and Sweatt, 2005). These result in a clear understanding of structural changes in both the pre-synaptic sensory cell and the postsynaptic motor cells, which in turn supports the modern notion of learning producing neuroanatomical modifications which underpin memory storage in a flexible, plastic, non-immutable way. It is partly what underlies the Neuropsychologists cheesy expression: whatever is fired often enough, eventually becomes wired.

The example previously discussed about simple single sensory-motor neuronal connections and memory making, is complicated many times over when thinking about the complexity of memory acquisition in hippocampal-dependent explicit memory. Indeed, as stated earlier this is better understood as a neural network of distributed processing and explicit memory.

One of the keys to understanding this is long-term potentiation (and its obverse: long-term depression). This is the most discussed and empirically supported mechanism of plasticity and memory storage, especially in the hippocampi and neocortex. The LTP model supports many of the necessary requirements for a robust theoretical approach in this area. It can accommodate speedy induction of memory traces, associate learning capability, short, medium, and crucially long-term duration. The reverse process of interference through specific pharmacological agents has demonstrated specific learning and memory impairments, which has garnered further support. In addition, equivalent manipulation (e.g. 5HTP introduction, a precursor to serotonin) which enhances LTP has also supported this model. Interestingly, it also through complex computer modelling, is perhaps the only model (and its specific rules) which can produce large capacity biologically realistic neuronal networks. Although complex, in essence LTP involves various combinations of amino-methyl-propionic acid (AMPA) and N-methyl-D-aspartate (NMDA) receptors for the frequently discussed neurotransmitter glutamate.

In summary, research at the epigenetic, cellular, neurochemical, and neuroanatomical levels, in particular over the last 20 years, all support the view of a brain that is highly functionally specialized cortically and subcortically, within which an inherently highly plastic neuronal system emerges. Moreover, this is also the best way of understanding memory, whilst acknowledging key areas such as the hippocampi, but not to the detriment of all the other areas that may be involved. In each of these potential systems the cortex is nearly always involved. It has been suggested that there are three primary memory systems, the hippocampal (including the para-hippocampal region) concerned with experiences and the self (relational), inference, essential for encoding and maintenance including recognition memory. The second is the emotional system based in the amygdala and with connections to and from the cortex. Finally, the neostriatal system (caudate, putamen, and ventral striatum) could be said to be central for sensory and motor performance and habit or procedural memory (Eichenbaum and Cohen, 2004). A final important point to note about memory, is that a defining feature, such as memory for space in the hippocampi is that it requires attention. Indeed, to some level or another attentional systems are as critical for memory systems as the hippocampi themselves. This would be especially true at the initial acquisition and encoding stage; but it does not exclude other stages.

Damage to each of these parts and the axons that bridge to and from them is common in acquired brain injuries and it is to these themes I shall now turn. A number of factors have been discussed in regard to potential damage in TBI. One of these is inflammation, to begin this is adaptive, but inflammation becomes problematic when it goes on too long (chronicity) and/or when too much inflammatory material is released at a point in time (acute). Hence, without a certain amount of inflammation the body will not heal, indeed it is an essential part of the healing process. Inflammation is also more important in some areas more than others. One obvious example is the chronic inflammation seen in the hippocampi in the aged population, the acquired brain injured population and specifically in the white matter tracts leading to and internally and away from this centrally important area of the brain and specifically memory functioning.

Even with milder TBI, axonal damage has been found consistently on autopsy. In particular, damage to the white matter output tract of the hippocampus; the fornix, was found to be damaged in 100% of those examined in a small but enlightening early study by Blumbergs et al. (1994). Numerous other studies of diffuse axonal injury in soldiers without direct impact, for example, blast shock wave injuries, have demonstrated clear memory deficits (e.g. Hoge et al., 2008). Further detailed understanding of this damage has demonstrated a loss of theta frequency power in the hippocampi after any TBI, which decreases with increasing severity, following this rat models have shown this can be reversed through direct stimulation, which then recovers memory behaviours (Lee et al., 2015). Wider distributed neural assemblies have now been able to be assessed through multi-unit recordings, which demonstrate disturbances in LTP in TBI populations (Tonegawa et al., 2015; Jaeger et al., 2018). Much is still to be fully conceptualized, such as the relationship between so called "place" cells in the hippocampi and contextual information outward for spatial and time understanding and how these networks may be specifically disrupted.

In addition, TBI axonal injury results in significant delays in synaptic inputs, these in turn effect adequate neuronal assemblies critical for encoding and episodic memory coordination. Finally, the kinds of axonal injury apparent in TBI disturb more distal connections between the hippocampi and places such as the pre-frontal cortex via the cingulate bundle. These ideas have been summarized by Wolf and Koch (2016), in their article on networks and TBI. This may be understood specifically for memory damage following TBI, but it should also be more widely understood as critical for recovery and neurorehabilitation from TBI as a whole, as Jassam et al. (2017) neatly summarizes: 'The key to developing effective therapies for TBI is to better understand and identify the precise mechanisms underlying TBI-related primary versus secondary pathology.'

It is something I focused on earlier, but in essence, I feel the longer-standing problems are primarily the result of these secondary injuries, apart from the axonal damage discussed. These develop over minutes, days, months, and as will be shown years. In large measure these can be divided into several potential pathways, including: excess glutamate, free radical damage, and neuro-inflammatory responses locally and systemically.

In an earlier study of inflammatory processes in TBI, Lenzlinger et al. (2004) a number of markers for inflammation were found in both CSF and serum (e.g. soluble interleukin-2). Whilst these peaked around the first week after the injury, they continued to be high over the following three weeks. One specific marker: β-microglobulin (β2m) is an interesting marker that appears in a number of different brain pathologies, associated with memory problems. It is now considered to be a systemic pro-ageing factor. However, it is also clearly implicated in cognitive decline, and specifically it is inversely related (in an almost dose-dependent manner) with hippocampal decline as the progression of Alzheimer's disease occurs. In other words, the greater the amount found in the hippocampal region, the further the decline and the later the stage of Alzheimer's is apparent.

In similar fashion, higher inflammatory agents in the brain have been demonstrated to show lower integrity of white matter in the brain (Bettcher, 2012). The reverse has been shown in uninjured, non-pathological, brains; most especially those people who exercised regularly whom had greater integrity of white matter and less inflammatory markers (numerous, but see Lee et al., 2017 or Torres et al., 2015, for helpful reviews).

This information moves us from the obvious anterograde amnesic presentation, with characteristic recall and recognition problems associated with medial temporal and diencephalic structural damage. It helps explain the wide-ranging and frequently counterintuitive (at least in terms of basic functional neuroanatomy) presentations of memory disturbance. For example, direct damage and the subsequent secondary damage in the frontal lobes, if the TBI is severe enough will produce different memory results than equal severity without significant frontal injury. Those with such damage will show search and verification deficits, memory organization, and self-initiated cueing disruptions, than those with comparatively less (Kim et al., 2009).

Taking these concepts together, the beginnings of a more rounded explanation of the complexity of TBI and its consequences can be suggested, especially for the memory assemblies, networks, and structures previously discussed. Thus studies have demonstrated that reactive microglia were found in 28% of cases up to 18 years after a TBI. Moreover, ongoing white matter debasement was also apparent. Not only this but the CC lost on average 25% of its thickness, one year post-TBI. The evidence accumulated to this point demonstrates white matter problems and ongoing neuroin-flammation years after the insult itself. The precise mechanisms apart from the traumatic effects directly on the axonal cytoskeleton, and later axonal degeneration, have been discussed but also involve excitotoxicity (e.g. over release of glutamate and aspartate), which also has related sodium, po-tassium, and calcium disturbances. Insults to any parts of the brain trigger astrocyte proliferation, these can result in a reactive state which promotes the intermingling of pro-inflammatory substrates such as microglia which eventually lead to a specific scar-like problem, which can then disrupt neuronal recovery and plasticity. In addition, mitochondrial dysfunction contributes significantly to metabolic and physiologic problems and even-tually to the death of cells. In a parallel problem impairments in natural clearing of cells through autophagy are also disturbed. There are also con-comitant disturbances in the electron transport chain and the catastrophic overproduction (and a comparable inability to manage) of reactive oxygen species. Finally, we return to the overproduction of pro-inflammatory cy-tokines, such as IL-1β, IL-6, and TNF-α, the greater the expression the more profound the severity of TBI. Whilst these have been demonstrated across many areas in TBI, as has been shown the hippocampal, para-hippocampal and related memory networks are most likely to be hit hardest through these mechanisms (Ng and Lee, 2019).

Part III

The combination of clockwise and counterclockwise vicious cycles

Chapter 7

Establishing the link between brain injury and sleep disturbance and sleep disturbance and brain injury

Sleep problems and brain injury make each other worse

Sleep disorders reported in acquired brain injury patients include but are not limited to: insomnia and hypersomnia syndromes, circadian rhythm disorders, excessive daytime sleepiness, delayed sleep phase syndrome, narcolepsy, and restless leg syndrome (RLS). Estimates of sleep disturbances vary in the brain-injured population. At even the lowest estimate the degree is significant (depending on the study, between 30% and 98%). With recent reviews suggesting that they are at least three times more common than in the general population, (Wickwire et al., 2016). Conservative estimates suggest at least 60% of the brain-injured population suffer from some form of sleep disturbance. In order to understand some of these issues more deeply two common types of acquired brain injury will be examined, TBI and then stroke and their interaction with sleep.

As previously discussed, sleep disturbances can increase depression, anxiety, fatigue, irritability, aggression, and one's general sense of well-being in the non-brain injured population. Irrespective of this, what we know is that TBI causes these kinds of problems to occur. In other words, organic damage can cause the problems that we have listed but sleep disturbance can also cause them. In many respects, what we would expect is a synergistic effect from the two different causes.

In a 2010 study, Shekleton et al. found that patients with acquired brain injury, and with raised anxiety and depressive scores had more disturbed sleep than healthy controls. They showed decreased sleep efficiency (SE) and increased wakefulness after sleep onset (WASO). As the book will explore later, the direction of causality may differ, and at the very least be more complex. It may well be the problem originates from sleep disturbances resulting in mood disorders, which then affect sleep further. This is complicated further by the knowledge that TBI patients show an increased tendency for all mood disorders. Shekleton et al. (2010), also found that slow-wave sleep (deep or NREM sleep) was increased across all traumatically brain-injured (TBI) patients. Melatonin is, as stated earlier, the provider of the window to sleep, its

DOI: 10.4324/9780429199066-7

waxing and waning is essential for a healthy circadian rhythm and it helps regulate the sleep-wake cycle. Its relevance to the circadian rhythm is very well established. Most tellingly, TBI patients had substantially depressed evening melatonin production, later (after six months) in their recovery. Melatonin level was significantly correlated with REM sleep but not SE or WASO. Shekleton et al. (2010), concluded that: ' ... evening melatonin production may indicate disruption to circadian regulation of melatonin synthesis.' The results suggest that apart from the significance of melatonin; elevated depression (and, to a lesser extent, anxiety) is associated with poorer quality of sleep. In addition: ' ... increased slow wave sleep is attributed to the effects of mechanical brain damage.' This was irrespective of melatonin levels and maybe an independent predictor of deficiencies in this phase of sleep. This last finding is very important as it conforms to severity and SWS problems in clinical practice.

Given what has been discussed it is therefore likely to expect greater sleep disturbance in the earlier days and weeks of recovery from any kind of brain injury but most especially a TBI due to the wide-ranging disturbances in neurotransmitter functions and production (e.g. massive increases in glutamate). To this end, a number of research efforts have looked at the natural history of recovery from such injuries. Makley et al. (2008) followed a group of closed head-injured patients during the subacute phase of their treatment. Overall, they identified 68% of these had disrupted sleep. In the study, they also investigated if sleep efficiency correlated with the duration of posttraumatic amnesia (PTA). Indeed, they found that those patients who had emerged from PTA displayed significantly better sleep efficiency scores than those that hadn't. In those with ongoing PTA, more specifically, they judged that for every 10 unit increase in a standard sleep efficiency score an increase of 1 unit on the O-Log (standard measure of PTA) was found. The conclusions were that not only was sleep disturbance common after a TBI, improvements in sleep efficiency correlated with resolution of PTA. Subsequent work by Sherer et al. (2009) and Nakase-Richardson et al. (2013) supports the linkage between sleep disturbance and PTA quite robustly. The latter found the:

> ... presence of moderate to severe SWCD (sleep wake cycle disturbance) at 1 month postinjury made significant contributions in predicting duration of posttraumatic amnesia ($P<.01$) and rehabilitation hospital length of stay ($P<.01$).

Later work by, amongst others, Gardani et al. (2015) found a number of key factors worthy of attention. First, unlike Nakase-Richardson et al. (2013) they noted the novel development of sleep problems or the worsening of existing problems post-injury. In other words, whereas previous work implied at least in some respects a bi-directionality of causation that runs from brain injury to resultant sleep problem and vice versa, later work has found better evidence for unidirectionality from brain injury through to sleep disturbance. The other key finding, known for many years through clinical impressions was that there is a high degree of chronicity to these problems. Ponsford et al. (2013) reached the same conclusions in regards to both aspects. In a reasonably sized matched sample study, they found that compared with non-injured controls those with a TBI had substantially poorer quantity and quality of numerous sleep parameters. Another finding was the chronicity of these problems, alongside a clear association with greater severity leading to greater total sleep time and increased daytime fatigue. Interestingly, this latter point has been debated for some time. An earlier study by Kempf et al. (2010) concluded that numerous studies have demonstrated injury severity is not a good predictor of the degree of sleep-wake disturbance. In 2019, El-Khatib et al. found that severity is only useful as a predictor of sleep disturbance when increased co-morbidities, longer hospital stay, and psychoactive medication were factored alongside it. In my own view, if one digs down into this paper, these co-morbidities are also more likely to have increased inflammatory problems, which as we know exacerbates and is also a product of sleep problems.

The debate now seems to have moved on to injury severity and type of sleep disturbance, rather than simply degree. In this regard, a small but informative study by Bogdanov and his colleagues in 2019 is instructive. They found that moderate (but not severe) TBI predicted excessive sleep, daytime sleepiness, sleep breathing disturbance, and greater overall sleep disturbance. In contradistinction, severe TBI (but not moderate) displayed significantly more disturbance in initiating and maintaining sleep.

Of equal validity in this more nuanced understanding is the study undertaken by Botchway et al. (2020). They followed up children with TBI into adulthood. SWD were investigated for prevalence; sleep quality, insomnia, and excessive daytime sleepiness. Standardized measures of severity together with standardized sleep measures were used. In addition, participants were compared to a typically developed sample. Interestingly, all types of severity did not differ in regards to insomnia or excessive daytime sleepiness. Of greater interest was how 20 years after their respective TBI, the severe group had less frequent SWD. Moreover, the moderate TBI group had the highest risk of SWD, especially overall sleep quality than the severe group. Anxiety and pain were associated across different severities with a higher risk of poor sleep quality. The latter was specifically associated with insomnia and excessive daytime sleepiness. Botchway and colleagues (2019) conclude that:

> Our findings indicate that sustaining TBI in childhood can increase risk of SWD in young adulthood, particularly following moderate TBI. Routine assessments and treatment of SWD, as well as anxiety and pain in children with TBI, should therefore continue into adulthood.

It is also apparent that something highly specific to TBI is occurring with sleep disturbances. When other conditions (e.g. chronic pain) or even traumatic spinal cord injury are examined they don't have the range, specificity, or type of sleep disturbances apparent in the TBI population (see for example: Wiseman-Hakes et al., 2016). In a study involving a very large number of subjects, Albrecht and Wickwire (2020) concluded:

> Individuals with TBI were more likely to have any newly-diagnosed sleep disorder before (14.1% vs 9.4%, $p < 0.001$) and after (22.7% vs 14.1%, $p < 0.001$) the index date. In fully adjusted DID models, TBI was associated with an increased risk of insomnia (rate ratio (RR) = 1.17; 95% confidence interval (CI) = 1.08–1.26) and any sleep disorder (RR = 1.13; 95% CI = 1.08–1.19).

As to the types of sleep-wake disturbance most apparent after a TBI, there are common themes. The most frequently encountered in clinical practice and research are insomnia and hypersomnia. These are then followed by sleep-disordered breathing, circadian rhythm disorder, parasomnia, movement disorder. Clearly fatigue is such a common finding after TBI it has to be included, but it has so far proved impossible to fully extract which aspects are as a result of the organic damage and which are sleep disorder related.

The natural course of sleep disorders in this population is unclear. Such disorders can develop during the immediate phase after the brain injury, later during inpatient neurorehabilitation or after some considerable time when the person is repatriated into a community setting. The reasons for this are an idiosyncratic combination of factors which range from the premorbid individual health, cognitive reserve, and general physical fitness of the individual, through to the severity of the TBI itself, the psychology of the person concerned, the context of recovery (e.g. very isolated, or in a supportive family) wider social and environmental issues (e.g. unnaturally dazzling LED lampposts, neighbourhood noise). Moreover, as has been shown repeatedly; the neurobiological aspects related not only to the severity but the specific way the body reacts to such an injury. All of which then have direct and indirect effects on the person's recovery through the interplay between these factors and potential co-morbid or developing problems such as pain, cognitive deficits, and emotional problems. These ultimately result in wide variations in overall quality of life. However, studies vary in their definitions of acute, subacute, and chronic. A useful, workmanlike definition will be used for this chapter. That is work prior to six (6) months post-injury will be defined as acute, work undertaken after six (6) months post-injury will be defined as chronic.

In spite of a wealth of anecdotal evidence from patients and clinicians, it has only been in the last ten years or so that these problems and their interplay have become more widely acknowledged. Indeed, it would be fair to say that in the UK, the author still finds it hard to access sleep services for his patients, frequently through either a lack of understanding in other professionals or simply (and as commonly) a lack of specialist sleep resources.

However, it feels at this point something is beginning to change, and there is a greater potential for understanding these problems and thus their amelioration. In this regard, it is not simply the clinical heft of opinion but certain key research efforts that have enabled us to view the problem more clearly. Having written this sentence I, rather depressingly, attempted to get a sleep assessment undertaken with one of my patients during the last week; only to be told by their GP, they only refer on for suspected sleep apnoea!

In any event, I would mark out a paper by Duclos and her colleagues as pivotal in regard to progress; it requires wider attention. Published in 2017 it outlines a small investigation of 30 patients who had a TBI (moderate through severe). Careful plotting of cognitive functioning (via the Rancho Los Amigos Scale) alongside wrist actigraphy measurements, although crude, revealed critical relationships between consciousness, sleep, and brain injury recovery.

Strong relationships were found which demonstrated improvements: ' … in consciousness and cognition in parallel with sleep-wake quality.' The conclusions of the research team were important in regards to the question of direction in causality. They stated:

> Our results showed that when the brain has not sufficiently recovered a certain level of consciousness, it is also unable to generate a 24-hour sleep-wake cycle and consolidated nighttime sleep. (Ducos et al., 2016)

Subsequent studies have elucidated these ideas. First, and fundamental is the work of Bigué et al. (2020) who have demonstrated the usefulness, sensitivity, specificity, and robustness of actigraphy as a valid measure in this domain of research. Second, in their helpful review: Fedele et al. (2020), together with one by Lowe et al. (2020) they demonstrated that sleep disturbance is particularly increased during post-traumatic amnesia. The kinds of disturbances found were also similar from these two separate reviews. In that, difficulty initiating sleep, fragmentation, poorer quality, and fundamental alterations in sleep architecture (e.g. REM) were the most frequently reported or measured.

The main issues with these research reviews were the frequent complaint of limited numbers of research papers, limited participant numbers, or limited numbers demonstrating the use of polysomnography in their work. The latter is perhaps less functionally justified when Bigué et al.'s work has been more widely disseminated and the clinical experience of working with the usually very distressed and disturbed TBI population is taken into account. If one has ever met a behaviourally disturbed person in the early days of their recovery, whilst in PTA, you'll understand what I'm driving at! A later study by Zeitzer et al. (2020) found that whilst polysomnography may be the gold standard, actigraphy is still useful but must be understood as a less sensitive proxy. Perhaps of most interest was this study's finding that overall actigraphy underestimated the level of sleep disruption when compared with polysomnography. However, I would reiterate, that if you've ever actually managed someone in PTA, with the practicalities of hooking them up to a full polysomnography kit, you'd soon realize the reason so few studies have taken place during this period of recovery. It is perhaps an invitation for the sleep specialist to join with the brain injury specialist to formulate a way forward.

Insomnia in TBI

Whilst one examines the literature on insomnia and TBI, differences emerge irrespective of severity, whilst others are highly dependent on severity. The classical diagnostic criteria for insomnia are most frequently met after a milder brain injury. The upper estimates of 70% described earlier seem to be more apparent here. Moreover, when digging deeper it would appear that the classical problem of falling asleep or delayed onset is the most frequent of these criteria followed by early waking (Ouellett et al., 2015), although in an early study of 'minor' brain injury and sleep it was nocturnal awakenings that predominated (Perlis, Arioles, and Giles: 1997). However, a highly particular population, of note is the potential for repetitive minor/mild brain injuries (Ne: concussions) to increase the likelihood of developing sleep problems. Bryan (2013) found that for military personnel, insomnia was slightly less than the general population on self-report measures (around 6%), but increased to 20% after a single mild TBI, through to 50% in those whom had experienced multiple minor TBI's. Obviously, the usual caveats apply in self-report studies, but it is illustrative and has been found in similar stepwise degrees in later studies (Spira et al., 2014).

Insomnias caused or exacerbated by other co-morbid conditions subsequent to a TBI are frequent. The common existence of post-traumatic depression and/or anxiety is much higher than in the general population. Prevalence estimates vary but range from 20% to 60%. A recent meta-analysis found anywhere between 36% and 50% prevalence for clinical anxiety disorders (Osborn, Mathias and Fairweather-Schmidt, 2016). Similar findings have been found for depression. In their most recent systematic review: Ryttersgaard et al. (2020), found depression was common. However, their estimated range was between 1.6% and 60%, which points to a variable study quality and lack of agreed terms of reference, in the studies examined. A more frequent finding was that both disorders were more prevalent in the younger population (below 50) and increased with TBI severity. The understanding being that whilst these disorders may not be causative, they are much more likely to be an exacerbating or at least maintaining factor in sleep problems. Of pertinence here is the view commonly held that the precipitating event may have been physical, such as the original TBI, which then corresponded to basic neurobiological changes which potentiate a mood problem. The subsequent maintaining factors could be different, such as anxiety, which commonly causes prolonged sleep onset (Staner, 2003).

Pain in TBI

Intimately linked, and more robustly studied is pain following TBI, especially that of headache. Nampiaparampil (2008) in his systematic review concluded: ' ... that chronic pain is a common complication of TBI, and was independent of psychological disorders, such as PTSD and depression.' He also noted that by far the commonest chronic pain problem was headache (found in 57.8% of those studied). Later studies and reviews have confirmed this form of chronic pain as persistent and unrelentingly linked to sleep problems (e.g. Ashina et al., 2019). In addition, Hou et al. (2013) found that post-traumatic headache suggested a greater risk of insomnia than the severity of index TBI. A later review by Grima, Ponsford and Pase (2017) found less conclusively in regards to the severity of the index TBI linkage, but as powerfully in regards to the commonality of chronic pain and, most especially, headache after a TBI and the clear correlation with sleep disorders. They also noted the difficulty in finding the direction of causality. However, the persistence of chronic headache could predict insomnia and vice versa depending on the study reviewed.

Hypersomnia (including excessive sleepiness and increased desire for sleep) in TBI

It is with hypersomnia that increasing severity of TBI leads to higher frequency and degree. It is interesting to note that most research has focused on the earlier stages of recovery. They have found that excessive sleepiness and the increased desire for sleep resolves in large portions of this population after one month post-injury. Longer studies crossing the divide between acute and chronic patient journeys suggest that this may take up to a year to resolve (Raikes and Schaefer, 2016 and Watson et al., 2007). Whilst these studies may be useful, they fly in the face of experience in community settings and other areas where chronic cases are more frequent. In these settings, hypersomnia is most characteristic of the clinical populations encountered. Having had experience in the acute, post-acute, and chronic phases of a persons recovery from a TBI, I can vouchsafe that the longer post-injury and the greater the severity of initial injury the more likely hypersomnia in either form will be encountered. It may be assumed that this is a base rate error and that the sample I have encountered is bound to be unusual. However, this experience is supported through large self-report surveys. The UK charity: Headway, produced a survey in 2019 confirming this view amongst its membership. In addition, actigraphy studies such as that undertaken by Imbach et al. (2016) found that in a population of chronically (six months post-injury) recovered TBI patients the majority needed one to two hours more sleep each night, compared to healthy controls. Given the earlier examination of the neurobiology of TBI and the complications of subsequent haemorrhage, they also found these sleep problems were exacerbated in those with the latter sequelae. Moreover, as shown the neurobiological mechanisms of the sleep-wake cycle are heavily dependent on the self-same neurotransmitters that go awry in TBI, the effects are thus profoundly problematic. What is more remarkable, is the small subsection of this population, yet to be fully examined in the research literature, who escape with little or no sleep-wake problems at all.

Clinical experience is supported by the majority of the research literature, in that to a great extent these problems seem to persist. Kempf et al. (2010) undertook a prospective longitudinal study of TBI patients. They followed 51 patients over three (3) years and utilized standard measures such as the ESS, the fatigue severity scale, together with an analysis of insomnia, depression, anxiety, insomnia, and that of post-traumatic hypersomnia. Sleep-wake disorders (SWD) were found in 72% of those examined at six (6) months, and 67% of patients at three (3) years. With some level of overlap, specifically they found in just under half the disorders appeared directly related to the trauma.

Once again, we see that the relationship with injury severity is not linear. Indeed, in this study post-traumatic hypersomnia was independent of TBI severity, localization, outcome, gender, age, and even psychiatric symptomatology. In contradistinction, however, those patients whom clearly met the formal diagnostic criteria for insomnia all had been diagnosed with depression (none had suffered from this prior to their TBI; as this would have excluded them from the study). In addition, overall, 45% of those who displayed SWD studied had no other explanation beyond their index trauma that could explain it. In other words, this study appeared to demonstrate that in large measure it was the TBI itself that directly caused the SWD, not any co-morbid, or pre-morbid confounds, for almost half the patients with SWD. In addition, they stated: 'Lastly, the prevalence of post-traumatic hypersomnia increased slightly to 27%, indicating that sleep need after TBI remains pathologically increased for a long period of time.'

This may well be the case. However, I would argue that it is just as likely that the restrictions in the environment; especially social and economic mean that in my experience most survivors feel compelled to return to ideas of normality which necessitate earning money to live or an uneasy part-time work existence, together with their own and others expectations of stopping napping in the afternoon, as they have apparently recovered from their original TBI. All of these constraints lead to a reduction or complete absence of the daily rest that previously enabled them to get through each day. It could equally be that 'sleep propensity' during the day has not faded at all and the 'unspecified' tiredness was always there and has as a consequence of the foregoing been unmasked; not become more prominent.

In any event, what we also know is that these difficulties continue to a greater degree than the general population many years after the index event (Hammond et al., 2019). Moreover, chronicity of sleep disturbance appeared, once again, to be unrelated to injury severity but was related to the development of post-traumatic headache and depression. Depression, as has been argued, is also a risk factor for sleep-wake disorders and is also related to injury severity. A recent study by Powell et al. (2019) found that those who had suffered a mild TBI were significantly more likely to report depressive symptomatology than those patients (on admission and discharge from a rehabilitation centre) who had sustained a moderate to severe TBI. Thus, whilst the development of sleep-wake disorders may be independent of injury severity, the development of depression was dependent on this and could then further contribute to the development of such sleep-wake disorders. These studies demonstrate the complicated multi-factorial development processes involved in sleep outcomes after brain injury. And once again these issues remain in the longer term (Martindale et al., 2017).

The conundrum of obstructive sleep apnoea (OSA) in TBI

As will be discussed further, the relationship between OSA and stroke is much better understood than its relationship with TBI. In studies of OSA and TBI it would appear to be a risk factor for sustaining a TBI (Young, Peppard and Gottlieb, 2002). OSA could also be related to the kinds of medications commonly prescribed post-injury (Nakase-Richardson et al., 2016) or indeed be simply related to the TBI itself and develop entirely post-injury, potentially as a consequence of such things like dysphagia or other neurological consequences of the TBI (Viola-Saltzman, and Musleh; 2015). The latter study examined 60 adult patients in a retrospective analysis. The majority underwent polysomnograms in addition to other standardized measures of sleep and mood. Of significance, given the foregoing, a full physical workup of the throat, chin, and temporomandibular joint found no abnormalities, as importantly none had sleep complaints pre-morbidly. Once again, although there was no consistent relationship with injury severity, there were several relationships previously discussed. First, insomnia was found in 25% of the sample, half of whom were highly correlated with depression scores, the other half were highly correlated with anxiety scores. The most common sleep disorder was hypersomnia (found in 50% of the whole sample), which wasn't related to either measure of psychological distress. However, it was related primarily to sleep apnoea. Unsurprisingly, perhaps, sleep apnoea has often been found to be of the central sleep apnoea (CSA) type. After this in terms of frequency, it was followed by the mixed type (one that begins as a central and ends as an obstructive), followed finally in frequency by OSA.

The more startling finding from these and other studies is that the frequency of any kind of apnoea is far greater than that in the general population. This seems to apply even when one considers the majority of TBI sufferers are relatively young when compared to the diagnostic average age for apnoea (Webster et al., 2001). One of the more profound results of such widespread apnoea in a brain-injured population is exacerbation of both the existing cognitive deficits apparent due to the brain injury but also the potential for poorer neuronal repair. As has already been argued, having apnoea without a brain injury leads to cognitive impairments through repeated hypoxic events and the concomitant issues around free radical release and subsequent glymphatic overload; just at the critical period when further damage and subsequent inflammatory responses are least helpful.

This latter concern is better understood in stroke patients, to which I now turn.

As previously demonstrated OSA is now considered an independent risk factor for stroke. It is also extremely common for those presenting with a stroke or transient ischaemic attack (TIA). Depending on the study the range of apnoea in stroke patients ranges from 30% to 72%. It would be safe to assume that some level of sleep-disordered breathing was apparent in around 50% of this population, which is in effect almost twice what has been found at the very top end of estimates for sleep-disordered breathing in the general population. The risk factors have been further stratified into the severity of the OSA (as measured in the standardized way through the apnoea-hypopnea index – AHI). Thus, the greater the frequency and severity of stroke will be related to the OSA index. This is particularly strong in older patients, who display an AHI of ≥25 (for a recent review see Jehan et al., 2018).

In an earlier review, Hermann and Bassetti (2009) identified that between 50% and 70% of stroke patients present with some form of sleep-disordered breathing. In contradistinction to those with TBI the majority are of the OSA type, followed by mixed then central. This largely depends on the type, extent, and severity of stroke. For example, bi-lateral strokes are more commonly associated with central sleep apnoea, whereas uni-lateral cortical strokes are more commonly found to have associated OSA in their risk profiles. In addition, OSA seems to be more apparent in the earlier period post-stroke, with some improvements but has been demonstrated to continue. Indeed, estimates suggest that AHI ≥10 will be found in patients two to three months after the onset of the index event (Rola et al., 2007). As previously examined, it would appear from the most recent research that OSA, is now considered an independent risk factor for stroke (with a much higher confidence than the same sleep disorders postulated causation in TBI) and it is also a *de novo* consequence of certain strokes. Moreover, even in the latter group where improvements are noted as part of the natural course of recovery, it is still found in a far greater percentage of the stroke population many months after the index event (Kim et al., 2019).

Hypersomnia in stroke

Although not as frequent as OSA, hypersomnia is still quite common in the stroke patient. Ferre et al. (2013) stated that between 1.1% and 27% of stroke patients suffer from this sleep-wake disorder. The wide variation in study findings was largely explained by differing methods of assessment and different time frames since the index event. The latter is profoundly important as whilst just over a quarter may experience this in the immediate aftermath, a recent study by Herron, Dijk, and Dean (2014) has thrown this into doubt and perhaps found a closer rate to that experienced by clinicians working in the community of 34%. Undoubtedly, part of the variation found is the simple disagreement over terminology. Some studies use extreme daytime sleepiness (EDS), others use hypersomnia, others use daytime fatigue, whilst yet more use the idea of excessive daytime fatigue (EDF). There is a large degree of overlap with these definitions but nothing, as yet completely agreed upon internationally. It is fair to say that working clinicians seem to make the difference cleave between hypersomnia and excessive daytime sleepiness which are synonymous; whereas daytime fatigue and excessive daytime fatigue which are largely independent of sleep and sleepiness. These ideas, as you will appreciate, makes for some problems occasionally in communication! What is much better understood is the profound consequences of such problems. As with untreated OSA, hypersomnia is an independent risk factor for stroke, (Boden-Albala et al., 2012). Obviously, the linkage between OSA and daytime fatigue in some studies means this is hard to disentangle easily. In addition, it is also much more common from the side effects of dopaminergic and noradrenergic medication (e.g. Bassetti et al., 1996). However, as with some TBI patients discussed there appears a subset of this clinical population who suffer from hypersomnia. A central hypersomnia is more commonly found in patients with damage to the reticular activating system (RAS). Strokes are especially problematic and commonly result in hypersomnia if the bilateral thalamic tracts are involved. However, anywhere that lesions occur in or around the ascending RAS, can cause this type of problem (Ferre et al., 2013).

Insomnia in stroke

If the common criteria discussed earlier for the diagnosis of insomnia is used a significant proportion of stroke survivors suffer from it. An early study found that 68% of those admitted to a stroke facility (post-stroke acute service) suffered from it and even after 18 months, almost half still did so (Palomäki et al., 2003). A later thoroughgoing review and meta-analysis conducted by Baylan et al. (2020), indicated a pooled prevalence (when understood using both non-diagnostic and diagnostic tools were combined) of 38.2%.

Depression and anxiety in acquired brain injury

More insomnia symptoms were present for those with some level of emotional or psychological co-morbidity. Indeed, as with TBI patients most studies have found that the significant factor in predicting sleep-wake disorders other than the highly specific neuroanatomical locations previously demonstrated, is depression (e.g. Suh et al., 2014). An early meta-analysis of depression amongst stroke survivors found that it had a prevalence in 29% of cases studied. A cumulative incidence figure of between 39% and 52% within five (5) years of stroke was found (Ayerbe et al., 2013). A later study by Mitchell et al. (2017) found similar rates; in that any severity of depressive disorder (mild, moderate, or severe) was found in 33.5% of cases studied. It was found to be more common in left hemisphere strokes, which is in contradistinction to the commonly held view of greater right hemisphere damage is associated with depression. In addition, dysphasia and either a personal (pre-morbid) or familial history of mood disorders also increased the relative risk of developing post-stroke depression. In their 2020 meta-analysis, Knapp et al. demonstrated equally higher rates than the general population for clinical anxiety, either through interview schedule studies (18.7%) or through studies using standardized rating scales (24.2%). Slightly higher rates were found in another recent study (Tang et al., 2020). This systematic review demonstrated that post-stroke depression peaked three to six months after the index event. When pooled, the results implied an average of around a third of post-stroke survivors developed depression and this only appeared to drop slightly some three years after the index event. It would thus be safe to conclude that depression and anxiety have a higher rate of incidence and prevalence post-stroke than in the general population. Moreover, the linkage between these disorders and sleep disorders is well established; although once again probably bi-directional in nature. It is helpful to think that depression and anxiety are implicated more strongly as time moves on, as critical maintenance factors, of a sleep disorder.

Pain in stroke

Similar to TBI a frequent finding post-stroke is pain. Depending on the study this can be as high as 70% of post-stroke patients experiencing pain on a daily basis (Klit et al., 2011). Of course, the reported prevalence differs depending on study design, definitions, and time since the index event. There is wide consensus amongst clinicians and overall study weight that pain is common after stroke. Frequently patients, especially in the first year attribute this entirely to their stroke. However, studies suggest that central post-stroke pain (CPSP) is the most frequent. This was followed in order of frequency: secondary pain from spasticity, shoulder pain, complex regional pains, and then headache (Harrison and Field, 2015). As will be discussed in further detail later, a great deal of evidence points to thalamic involvement in CPSP, and pain problems more generally. CPSP was first described by Dejerine and Roussy (1906) as: 'syndrome thalamique,' and has been known as such ever since. Subsequent work, especially over the last 20 years, primarily due to advances in functional MRI techniques, has widened our understanding of these pain problems and their neuroanatomical substrates. The ventral posterolateral nucleus (VPL) and ventral posteromedial nucleus (VPM) structures in the thalamus have been repeatedly implicated in CPSP development (e.g. Treister et al., 2017). Other tracts, such as the trigeminothalamic, in addition to parietal cortex and wider post-central gyrus locations, have also been implicated in CPSP development at a later point in recovery, during the first six months following the index stroke (Nakazato et al., 2004). Nociceptive pain associated with regional pain, shoulder pain and spasticity, is common and results from the physical damage to the part of the body concerned. That is unlike headache, which although common, is not as clearly understood. The most likely underlying mechanism is the trigeminothalamic pathway and surrounding vascular structures, although this is still tentative (Sommerfeld and Welmer, 2012). A much more recent extensive systematic review and meta-analysis by Harriott, Karakaya, and Ayata (2020) found that headache occurred for between 6% and 44% of the ischaemic stroke population. Moreover, in this analysis posterior circulation and females had a significantly greater odds ratio risk of developing a headache post-stroke.

As can be seen, the interplay between TBI or stroke and sleep disorders is complex. Whilst both brain injuries have been implicated in a range of common sleep disorders; both from and subsequent to them, the exact relationship is not always fully explicable and predictable. Moreover, these difficulties are compounded by the common co-morbidities of depression, anxiety, and pain.

Summary: the beginning of joined-up thinking

With the potential exception of OSA (and slightly less robustly hypersomnia) and stroke, no other form of sleep disorder can be pinpointed as an independent risk factor for an acquired brain injury. It is true that having a sleep disorder increases the risk of such; the findings are at the level of a potential exacerbating factor and not a clear linear relationship.

Any form of acquired brain injury will increase the risk of sleep disturbance. This may be as high as 60–75% of those who have suffered such an injury. This is especially true for the acute recovery phase, where repeated clinical experiences and studies have demonstrated extremely high sleep disturbance in the first six months of recovery. These findings have very clear implications for assessment and the current parlous state of dedicated sleep treatments for this population, at any stage in their recovery. This is critical, as overwhelmingly the evidence is that for every quantitative, but especially qualitative improvement in sleep, there is a concomitant improvement in some measure of brain injury recovery or performance (e.g. PTA resolution, later mood measures, or memory performance).

Severity of TBI appears to promote certain types of sleep disorders over others. Sleep disorders are most prominent for milder TBI's in toto. People with moderate TBI's have sleep disturbance less commonly chronically. However, they are more likely to have them than severe TBI patients; found in long-term follow up. As a general rule of thumb, it is also largely true to say that milder injuries have a greater propensity to produce a classical insomnia type, and moderate through severe injuries are more likely to produce problems associated with hypersomnia and other sleep disorders. This is particularly true if chronicity is taken into account, as although increased numbers overall recover better sleep function, the residual population of those who sustained a severe TBI will more likely continue to suffer hypersomnia years after the event, which is demonstrated both through research data and clinical experience.

This may appear counter-intuitive until you pay attention to elements beyond the severity of the brain injury. Severity alone, although helpful as a potential explanatory mechanism, especially for the findings of insomnia and hypersomnia previously noted, needs the additional support of other factors to explain overall variance. In effect, if the patient has pain, depression, anxiety, and to a lesser extent other co-morbidities previously discussed, any residual sleep problems are magnified beyond a simple one-to-one relationship with severity. Moreover, at least one form of pain; headache, is generally considered a more significant risk factor for a specific form of sleep disorder: insomnia, than severity of TBI. Whilst headache is an important consideration impacting upon sleep, especially for women who have suffered a posterior stroke, other pain problems are equally important to consider. Most critically, thalamic pain syndromes are associated with thalamic and para thalamic strokes (inc. bleeds). It takes a very small step to then confront the central importance of these latter areas for sleep and memory functions. Put simply the thalamus is central for NREM sleep, which is of course supported through melatonin release and its advancement of spindles. Related structures, such as the hypothalamus, control aspects of temperature, sleep, and arousal, which may be familiar to the careful reader.

Stroke differs from this to a degree and is especially sensitive to the neuroanatomical substrate that is damaged. In this regard, there may well be a sleep disorder apparent after some "simple" cortical strokes, but after a bi-thalamic stroke it is almost inevitable; likewise for a severe grade sub-arachnoid haemorrhage. The focus here needs to be on the neuroanatomy as well as the time from the point of the index event.

Other factors are also strongly associated with stroke and sleep disturbance. Overall, around a third of stroke patients suffer some form of mood disturbance. Most prominently in the first three months this is likely to be anxiety. Depression seems to peak between three and six months post-stroke. However, as previously discussed it is only some three years after the stroke event that depression appears to decline somewhat. It is probably more helpful to view chronic depression and anxiety disorders as maintenance factors for the basis of assessment and treatment, rather than initiating factors. That is unless, of course, these were part of the patient's pre-morbid history.

In regards, to at least the various mood and related disturbances, it is fair to say that a large proportion of these begin with whatever type of brain injury (and are thus not pre-morbidly apparent), but that over time they may come to be better understood as a maintaining factor in sleep disturbance. If we then re-visit earlier chapters on heterotopic ossification and memory, the general idea may be summarized as some form of premature ageing in the acquired brain injury patient. There are considerable parallels with the older adult neurobiological profile and outward behaviour. To take two examples: first, inflammatory processes become chronically apparent in both populations, leading to other commonalities in the internal world. Secondly, the need for ongoing daytime naps is increased alongside disturbances in particular stages of the sleep cycle – most apparent in deep (SWS: N2 and N3) stages along with alterations in REM. Of course, as we have seen the former, in particular through disturbances in the glymphatic system will mean further increases in problematic inflammatory changes. The system is unable to fully clear the daily accumulation of waste products in addition, to the already huge amounts of increased waste products (e.g. superfluous glutamate), which will then lead to even further decline. In some respect, the problem of brain injury and sleep disturbance altogether can be crudely summarized as the waste that brain can't wash away. In other words, at least one part of this problem and the presentation of those who suffer from both is not simply a damage problem but a: *waste disposal* problem.

The main difference here is the expectations of the self and others in these different age groups. By this I mean it is psychologically and socially more acceptable for the older age groups to nap in the day and report difficulties with sleep as the ageing process gathers pace. However, it is largely held as unacceptable for these same behaviours to be displayed or reported in younger age groups. These behaviours, especially napping in the daytime, would likely be met with disapproval, even opprobrium, potentially connoted with laziness. Thus in large measure, the younger age groups treated in clinical practice feel compelled to return to work, on occasion far too early in their recovery. On several occasions in the author's experience with too great a weight of internal and external expectations based on unrealistic assessments from those with little or usually no experience of acquired brain injury.

Heterotopic ossification was discussed earlier as a significant problem for those surviving a TBI and stroke. In regards to sleep, it may surprise the reader to note it has also been shown to be a direct consequence of sleep problems. These sleep deprivation effects have been demonstrated across species from rats, through primates and more recently humans (predominantly women). From the study by Xu et al. (2016) which found significant differences on autopsy of rats in a large chronic sleep deprivation group, as opposed to the control group. They concluded that chronic sleep deprivation significantly reduces bone mineral density (BMD), as well as another marker of bone health which deteriorates bone microarchitecture. In addition, overall, it decreases bone formation and bone reabsorption markers. A study the following year by Lucassen and her colleagues (2017) which examined a large cohort (915 participants), found similar difficulties, utilizing dual x-ray absorptiometry (DXA) scans. Bone density, relative muscle mass, osteopenia, and sarcopenia, after adjustments for confounding factors, were found to be significantly depleted in those who had a delayed sleep timing (not necessarily standard sleep delay after going to bed) and decreased sleep quality. These associations were slightly stronger for women. In an even larger study of 11,084 postmenopausal women published in 2020, similar associations were found. Ochs-Balcom et al. 2020 performed a cross-sectional study of quality and quantity of sleep with whole body, total hip, femoral neck, and spine BMD. As with the earlier human study they used standardized sleep quantity and quality measures measured against DXA, bone mass, osteoporosis models. When all potential confounds available (e.g. alcohol use, physical problems, medication use, age, and so forth) were adjusted for, a number of highly significant findings emerged. Women who slept five (5) hours or less per night had significantly lower BMD in all four sites compared with women who slept seven (7) hours or more each night. The same sleep-deprived group also had a higher odds ratio of osteoporosis of the hip. The studies have reached powerfully similar conclusions over the last five years. Sleep deprivation in particular and poorer sleep quality both significantly impact upon bone health and for women, are much more likely to lead to osteoporosis post-menopause than non-sleep deprived postmenopausal women of similar age and background.

In regard to memory, I hope that earlier chapters demonstrated why memory is disturbed after TBI and other acquired brain injuries. A more recent trend in the sleep literature has been to understand the specific mechanisms underlying potential memory disorders leading directly from sleep deprivation and sleep quality problems. Although animal studies have examined these issues for considerably longer, it is only recently they have been supported by human research. Indeed, as long ago as 1924, Jenkins and Dallenbach were observing linkages between sleep and memory function. However, so many improvements in imaging (including fMRI and PET) have accelerated our understanding. This has occurred in tandem with advances in neurochemical theory and assays supporting complimentary models.

An early example of these kinds of studies was that of Noh et al. (2012). The team looked at relationships between hippocampal volume and cognition in chronic insomnia. They looked at a sample of previously diagnosed insomniacs and a matched non-insomniac control group. MRI scanning was undertaken in all participants and then hippocampal volume was measured manually. In addition, nighttime polysomnography and neuropsychological testing were applied to all equally. All these measures were then evaluated. Compared to controls, the insomniac group, unsurprisingly, displayed increased sleep latency and higher arousal indexes, and a decreased percentage of REM sleep during polysomnography. In addition, this group also displayed impaired verbal and visual memory and frontal lobe functional deficits. These elements were also related to decreases in hippocampal volume. Most of the clearest correlations were found between specific impairments (e.g. free recall in verbal memory) and reduced left hippocampal volume.

A review by Kreutzmann et al. (2015) demonstrated that there is a wide consensus that optimal sleep quantity and quality benefit neuronal plasticity. Later, they provide valuable evidence from a range of sources that the obverse also stands: sleep deprivation, results in impairments in learning and memory. These have been further specified as being highly dependent on the hippocampal and para-hippocampal systems; which sleep problems effect most obviously. A number of neurochemical substrates have been suggested as being affected by these problems. Kreutzmann and his colleagues concentrate on the potential for some familiar ones, as described in earlier chapters to be most affected, they go on to suggest:

> ... available evidence suggests that SD may impair hippocampal neuronal plasticity and memory processes by attenuating intracellular cyclic adenosine monophosphate (cAMP)-protein kinase A (PKA) signalling which may lead to alterations in cAMP response element binding protein (CREB)-mediated gene transcription, neurotrophic signalling, and glutamate receptor expression.

Now of course part of what has been discussed throughout this book is the importance of glutamate (amongst others) for a great many functions. It is crucial for memory performance but at the correct level. The review paper (*ibid*) goes on to conclude that the available evidence supports the view that when sleep restriction becomes a chronic condition, it reduces hippocampal cell neurogenesis and eventually leads to inevitable hippocampal volume reductions. It is at this point that the reductions in volume, alongside the reasons for that, lead to cognitive problems and an exacerbation of various psychological problems such as depression.

Finally, I will highlight a much more recent review by Cousins and Fernandez (2019). They argue that sleep is not only important for consolidating new memories; it is critical for all the stages of memory, and other aspects of memory processing. In the order of things; newly formed memories, with concurrently weaker strength of traces and connections between neurones are more fragile. As consolidation progresses, as I discussed earlier, the hippocampus becomes less critical and the para-hippocampal structures and then wider areas (e.g. frontal regions); as final transformation along with synaptic consolidation occurs at cellular levels. All these work in complimentary ways. This process, depending on the type and nature of the memory concerned may take many years to finally attain. A number of ways of assessing disruptions in sleep have been proposed. The common sleep – no sleep approach, was the early experimental norm. Then fractionated approaches in sleep experiments in the 1970s began to become pre-eminent. For example, the manipulation of deprivation of NREM or REM stages, which basically deprived either the first or last 3 to 4 hours of sleep during a night.

Some examples of the different aspects of memory problems in a commonly known sequence after sleep deprivation are needed. The most obvious problem is encoding, which has been demonstrated to be affected by sleep deprivation. Indeed, bluntly, there is a clear linear relationship between hours lost and the amount of material remembered. However, the specific relationship to specific encoding deficits is harder to pin down. Interestingly, even if some of the information presented has been remembered, it is usually partial, depending on the amount of deprivation. Experiments by Harrison and Horne (2000), showed faces were remembered but the context (critical for fluid memory processing) was not. Similar effects have been shown for auditory-verbal memory, visual memory, and even film clips (Saletin et al., 2016). As has been described earlier, even if the sleep deprivation is "overly rectified" through attempts to make up for lost ground on subsequent nights, no similar catch-up in memory performance occurs. It has thus been cemented as a problem of encoding and in addition, a problem that needs a good night's sleep on the first night after the new material has been learned. These ideas have been further examined through emotional and neutral presentations, and whilst some improvements occurred when emotions were paired, there was still a considerable decline in performance. The hypothesis is that context-dependent memories are highly related to hippocampal performance. Whereas emotional memories rely on the amygdala and prefrontal cortex, and as their distribution is wider they allow for slightly better remembering (Dolcos et al., 2005).

It would also appear that in acute sleep deprivation of REM, with N2 and N3 largely intact (the first split half preserved) encoding is unaffected. However, chronic sleep deprivation, using the same method does produce encoding problems (Cousins, et al., 2019), which again points to the power of chronicity and the implications for increased inflammatory problems accumulating over time.

Overall, it seems it is consistent that hippocampal and parahippocampal regions dominate in the encoding process. That is unless emotional content is paired, and with that wider recruitment occurs from other regions (e.g. the amygdala). It is also reasonable to suggest that SWS, especially (N3 predominantly) is crucial but not sufficient for this process, as demonstrated through shorter (one or two nights at most) sleep deprivation. However, as sleep deprivation becomes chronic, sleep spindle activity is reduced and together with other associated problems (increased inflammation?) the particular aspects of N3 sleep seem to recede in importance and a whole stages problem begins (Poh and Chee, 2017).

It is with the consolidation of memory that there emerges an even clearer understanding of sleep disturbances effects. Crudely, it appears that N3 (NREM slow-wave sleep) in particular, together with some aspects of REM are critical for this process of memory consolidation (Giuditta et al., 2014). Many total sleep deprivation studies have supported these general principles, some assessing long-term consolidation after six (6) months (see for example: Gais et al., 2007). Interestingly, and unlike encoding studies, it has been shown that some form of recovery sleep can produce some (not full) recovery of those losses to consolidation (Rauchs et al., 2008).

Moreover, N3 SWS seems to be important for consolidation of memory without necessarily any emotional content and REM sleep the reverse; it is crucial for consolidation of memory with emotional content. Neuroanatomically, it has been repeatedly shown that as memories are further consolidated, firing is shifted from purely hippocampal regions to wider neocortical sites. This occurs over lengthy time frames and according to Diekelmann and Born (2010), it is sleep that is the *fundamental driver* of this memory process.

Retrieval is largely defined as free, cued, or under-recognition formats, within each of these functional domains, different brain regions and neural networks are more prominent. In large measure, a number of these networks are connected in some (even marginal) way with the frontal cortex and are far less dependent on hippocampal processes. Interestingly, in each of these formats sleep deprivation appears to have smaller effects than for either encoding or consolidation. Lim and Dinges (2010) offer a wide-ranging review of the cognitive systems effected by short-term sleep deprivation and there was a clear robustness found in retrieval mechanisms. One could speculate that the lack of hippocampal involvement together with the more obvious deeper and stronger synaptic connections (a great many firings and thus wirings during the encoding and consolidation phases) allows for this robustness, but these hypotheses have not been adequately explored at the current time to be certain.

Finally, what we also know is that chronic sleep deprivation leads to significant increases in neuroinflammatory markers. Aboul-Ezz et al. 2020 found increases in a number of those (e.g. TNF-α, IL-1β, and IL-6) as well as a reduction in growth-stimulating hormone levels. Moreover, further assays revealed an increase in hyperexcitability (through increased glutamate, glutamine, and aspartate) and concomitant decreases in GABA, specifically during REM deprivation. As we also know, during sleep GABA in the VLPO inhibits histamine neurons, thus initiating further problems in associative systems and overall increasing excitatory problems; without the necessary breaking system fully operational. All of this recent research is in line with previous discussions in this book that point to similar disturbances in these same markers across both the pathologies under discussion. Indeed, as has been pointed out glutamate alone is released between cells, after injury, and as the most common excitatory one in the brain, it has been estimated to be involved in over half of all synaptic action potentials (see for example: Purves 2011), this release is profound. Glutamate is critical for information processing and essential fast signalling. There are also widespread glutamate receptors found throughout the cortex, very densely in the hippocampi, even in glial cells. It is also inextricably linked to nearly all other neurotransmitter systems (along with aspartate). Too little slows interneuronal communication, whereas too much is basically neurotoxic (Meldrum, 2000, and later Guerriero, Giza and Rotenberg, 2015). It is no wonder sleep difficulties add to an already massively overburdened system. It also suggests that potentially a treatment or combination of treatments may ameliorate both pathologies.

Chapter 8

Integration of thoughts

Can healing sleep problems heal brain injury problems?

If we then integrate the first chapters, we can then discuss the possible mechanisms of specific sleep disturbances for those with an acquired brain injury. In addition, to sleep and its effects on those people with a brain injury. There are clear implications for specific patient populations. An example being how frequently frontal contusions are found in the TBI group. Logically this would potentially affect the initiation of NREM sleep. This may be borne out by the literature, but it is more likely it has been insufficiently recognized. Moreover, and it is here that I think this gets even more interesting, both the sleep disturbance caused by the brain injury and the more direct effects of the brain injury potentially limit (or may indeed propel, if tackled) the neuro-rehabilitative efforts of the professionals around the brain-injured person. For example, if the damage was mainly sustained in areas in or connected with the temporal lobes, especially the hippocampi, will the memory deficits found commonly in TBI be exacerbated by the sleep disturbance, especially if it is of a particular kind of sleep disturbance? In addition, will the alteration or improvements in sleep disturbance enhance the recovery or even functional outcome of those same memory disorders?

The most obvious parameter that we would be able to modify is the overall quantity of sleep. However, I would argue that we should focus on the qualitative, as well as the quantitative aspects of sleep. It may well be more useful to look at enhancing a particular cycle of sleep related to a particular function associated with brain injury. Even more specifically it might not necessarily be related to a cycle that may best enhance outcomes, it could be the timing of the delivery of a hormone (e.g. melatonin). Or indeed, it may well be that certain key sleep interventions may be foundations to better quality sleep across all the disorders mentioned and the parameters under discussion around brain injury recovery. Thus, within the broad compass of sleep, there is a potential for a vast array of different forms of treatment.

DOI: 10.4324/9780429199066-8

To begin with the exploration of the neuroanatomy, and neurobiology of sleep-wake and brain injury, previously undertaken, will be brought together. As has been shown there are so many important shared neurobiological and neuroanatomical mechanisms in both these that we could mine new pathways to neuro-rehabilitation successes that have so far remained elusive.

Recalling the initial examination of the balance between wakeful and somnolent neurotransmitters, we became aware of the deep through higher structures involved: from brain stem, hypothalamic, basal forebrain, and superior frontal cortical structures in arousal. Inherent in these structures are the wide variety of neurochemicals, including, but not limited to histamine, serotonin, acetylcholine, orexin, noradrenaline, dopamine, and importantly glutamate, as activators. They play, at different stages, a beautifully conducted number of orchestras that lead to wakefulness, arousal, and attention. Recall also the reverse neuroanatomical progress from superior cortical structures to deeper brain stem regions to induce sleep, the modification of those self-same neurochemicals. In addition, consider the more somnolent ones such as melatonin and GABA. One begins to appreciate that in certain key structures and disturbed neurotransmitter functions, one may reasonably ask how could brain injury (especially TBI) not disturb sleep. It is to these commonalities that I now turn.

Neuroanatomy and neurochemistry

From the base to the top

Progress over the last decade in understanding the neuropathology of brain injuries for sleep-wake disorders has improved. However, as early as 1971, Crompton found that brain stem lesions were prominent in a great proportion of fatal TBI patients. Later studies supported this view, especially in those patients who died or were comatose. So called "radical disturbances" in consciousness occur in patients with these lesions. Thus, Parvizi and Damasio (2003) studied patients with brainstem stroke. They stated: ' ... in patients who had coma (n = 9), the lesions in the tegmentum were mostly bilateral (n = 7) and were located either in the pons alone (n = 4) or in the upper pons and the midbrain (n = 5).' They finally concluded that brainstem lesions are most likely to be involved in loss of consciousness. This also is consistent with the clinical experience of, in particular, midbrain and hypothalamic damage leading to hypersomnolent states.

These kind of considerations also make further sense regarding white matter damage in TBI and the classical caudal-rostral route to wakefulness, and the reverse to somnolence. Central to this conception is the interrelationship of the locus coeruleus in its wake-promoting (noradrenaline) power alongside the GABA, galanin, and subsequent disturbed adenosine production cycle somnolence promoting power. This has been elegantly discussed in their paper on general anaesthesia, sleep, and coma by Brown, Lydic, and Schiff in *The New England Journal of Medicine*, in 2010. In their review they suggest that:

> During the awake state, the locus coeruleus provides noradrenaline-mediated inhibition of the ventrolateral preoptic nucleus in the hypothalamus. Therefore, GABAA-mediated and galanin-mediated inhibition of the ascending arousal circuits by the ventrolateral preoptic nucleus is inhibited and the awake state is promoted. Adenosine, one of the brain's principal somnogens, accumulates from degradation of adenosine triphosphate during prolonged intervals of wakefulness... Adenosine binding and inhibition of the locus ceruleus lead to activation of the ventrolateral preoptic nucleus, which inhibits the ascending arousal circuits and promotes non-REM sleep.

The relationship between this somnolent and sleep seems clear but gets complicated through TBI and other acquired brain injuries. In this regard, numerous studies have found that adenosine is produced at far higher quantities than usual. This does not simply affect the obvious fatigue reports of brain injury survivors, as adenosine and its metabolites are implicated in a wide range of cellular functions. Other than its accretion during wakefulness, it has direct effects on adenosine triphosphate (ATP) and later cyclic adenosine monophosphate (cAMP) mechanisms. The latter, although primarily a messenger, can become overactive after brain injury and has been implicated in migraine, through over activation of the trigemino-cervical system. As shown earlier, part of the adenosine build-up is to do with initial reliance on local anaerobic energy production; producing a tenth of the energy in the aerobic electron transport chain, leading to a reduction of ATP units. This leads to increased adenosine waste and lactic acid build-up inducing greater imbalance in the sleep-wake cycle along with effects on the fatigue experienced by those who have sustained an acquired brain injury. This is without the further damage caused through reactive oxygen species, oxygen damage itself, pro-inflammatory problems associated with excessive glutamate and reductions in nitrous oxide production. Finally, local and para local oedema are brought about. It is thought to be part of the possible reasons for the chronic fatigue so prevalent in TBI populations. Moreover, whilst caffeine in the acute setting has been found to produce worse outcomes, caffeine in the chronic brain-injured populations has shown some benefit. The hypotheses for administration in both settings were the known adenosine receptor blocking action of caffeine; and given the large amounts of excess adenosine produced after a brain injury. This could mean the message about caffeine consumption and sleep needs slight modification for the brain-injured population (see Lusardi, 2009, for an early review). As part of the sleep process S, it, alongside other variables, may be modified to some effect. There may also be positive ways to reduce or stimulate this important neuromodulator. It is easy to see why stimulants such as Methylphenidate have been utilized to assist hypersomnia and fatigue states.

The further we travel up neuroanatomically the closer we get to the midbrain and the pineal gland; the centre of melatonin production, which is also extensively produced in the gut. After some early spikes, the production of this essential sleep hormone has been found to be consistently reduced after a TBI of any severity. Correspondingly, its regulation throughout the circadian cycle is disturbed (Grima, et al., 2016). The latter tested a small sample of patients with severe TBI, against a non-injured control group. The TBI group had on average a delay of melatonin secretion of approximately one and half hours (1 hour and 30 minutes) and potentially more disabling, a 42% decrease in overnight production. These effects, especially the significantly lowered production, have also been shown to last a considerable time after the event occurred. Shekleton et al. (2010) found that this lowering lasted at least one year post-injury. It should be borne in mind that the exact relationship between pineal gland damage and these effects was not directly tested, rather in most of these studies various proxy or analogous measures were used to understand the functional consequences.

In the shorter term, melatonin is potentially seen in reverse of these findings. Certainly, in the paediatric population it would appear that in the very early critical illness and intensive care stage, melatonin levels are extremely high. Marseglia et al. (2017), studied serum levels, rather than the more common saliva samples, of melatonin taken at frequent intervals. It contrasted healthy children, critically ill (non-brain injured) with children who sustained a severe TBI. They discovered a slope from healthy controls through critically ill children, which then became even more significant in the children with a TBI. I would suggest that some potential confounds were the medication the children were on, apart from this and the small sample size it builds on other research that has produced similar results. A study by Yan and Zhang (2019) found a similar slope with increasing severity in preterm infants with a brain injury. They found that infants with a mild brain injury had elevated serum melatonin (along with αII spectrin cleavage products: SBDP's), and that those with the severest brain injuries had the highest levels. They concluded that these results taken with the other research available suggest that serum melatonin levels, when taken early enough, are not only high they can be stratified for diagnostic purposes or as they state in their conclusion: 'Detecting melatonin and SBDP's has clinical value in diagnosing and assessing the severity of brain injury in preterm infants.'

As yet no study has adequately tracked the potential increased anti-inflammatory release of melatonin during these earlier periods after initial brain trauma through immediate post-acute and then the very long-term reduced levels seen in numerous other studied longitudinally.

Moving upwards again, a cluster of critical brain regions is found anterior and superior to the midbrain. These areas include, but are not limited to, the hypothalamus (and structures within this e.g. tuberomammillary nucleus), SCN, and basal forebrain. These areas are less well understood, as historically, the brainstem regions previously discussed have been examined extensively in TBI patients along with sleep disturbances, since the 1930s (Courville) and later through the ground-breaking work of Adams (1982, 1989, 1991, 1993, etc.) and even later still Baumann (2012).

Our ability to understand the subtleties involved in these regions has been enhanced through more modern imaging techniques, along with new knowledge of some neuromodulators such as orexin.

To this end the hypothalamus will be studied next. This small yet deeply differentiated (functionally) area anterior to the midbrain is of profound importance. In early studies, it was found that almost 80% of those who subsequently died from a severe TBI had brainstem and hypothalamic injuries. Of these the largest portion suffered more significant hypothalamic damage if raised intracranial pressure accompanied by herniation had occurred (e.g. Shukla et al., 2007), with anterior hypothalamic damage being predictive of shorter survival time than posterior damage. As demonstrated earlier whilst there is quite a heterogeneous pattern of sleep disorders following TBI, with no discreet relationships consistently found with location site, there are proxies that help make some sense of these disorders. The work of Baumann and Basetti (2005) has been instructive in regards to orexin production. In an early study, they demonstrated that 90% of the moderate to severe TBI patients they studied there were decreased levels of orexin. Whilst the majority recovered to functional levels there remained a significant minority with persistent problems with issues like narcolepsy and hypersomnolent states being the two most prevalent. In a later study by Valko et al. (2016) 58% of the patients studied who had sustained a TBI had clear neuropathological abnormalities of the hypothalamus. This same study found support for dysregulated orexin but even more support for long-term problems in anterior structures, together with damage to the tuberomammillary bodies of the hypothalamus associated with histaminergic neurons. These bodies are the only part of the brain that produce histamine. High quantities are made during the day and less at night. In addition, extracellular histamine levels in the frontal cortex are positively correlated with the amount of wakefulness (Chu et al., 2004). Higher levels of histamine have also been shown to be important in inflammatory responses, amongst other key functions. Moreover, the dense axonal projections from these bodies to the cortex have not only important histaminergic arousal properties; the bodies themselves are fundamental to learning and memory.

Another critical neuroanatomical structure that is not only important for wakefulness but also important for the cortical rhythms associated with cognition itself is the basal forebrain. This area is more rostral to the hypothalamus and is more frequently damaged, either directly through the trauma of traumatic brain injury, the peri-axonal damage around it, or through the neurobiological and neurochemical assault described previously in TBI. Having said this, its full function is yet to be completely understood. What we do know is that the degree of damage to this area is associated with greater severity of the depletion of the wake-promoting cholinergic neurones and correspondingly poorer cognitive performance. This is true of specific lesions associated with psychosis, dementia, and more pertinently TBI (Salmond et al., 2005).

Early studies, involving lesion examination, animal experimentation, electrical stimulation, and neuropharmacological administration appeared to demonstrate the basal forebrain as extremely important for increasing behavioural arousal and reducing overall sleep; especially that of deep (stages N2 and N3) sleep. Thus, much research effort seemed to conclude the basal forebrain was a fundamental part of the wakefulness component of the overall sleep-wake system (McGinty and Sterman, 1968 and later Szymusiak and McGinty, 1986). However, later reassessments and different research groups discovered the reverse. It appeared lesions demonstrated that the basal forebrain was important for the somnolent-promoting aspects of the sleep-wake system. Indeed, inactivation of the basal forebrain in these studies found large increases in delta wave activity, decreased behavioural arousal, and concomitant increases in deep sleep (for a review see Fuller et al., 2011).

These seemingly contradictory findings have only been understood as work has progressed in the knowledge of the basal forebrains three major neuronal types. These are the cholinergic, glutaminergic, and GABAergic ones. As discussed previously cholinergic neurones are very active during wakefulness and REM sleep, and completely silent during NREM sleep. Not only this: they are fundamental for the three basic pre-cognitive and cognitive systems: arousal, attention, and memory. Moreover, the gluta-minergic neurones appear to be the origin of these effects. Glutamate has strong excitatory wake-promoting effects and seems to act locally in the basal forebrain to excite cholinergic neurones and thus promote wakefulness and cognition. It is also apparent that these same glutaminergic neurones cause some specific GABAergic activation, the latter is known as a PV+ GABAergic neurone and unusually this also promotes wakefulness.

According to Xu et al. (2015) part of the problem of seemingly functional contradictions was found to be due to spatially intermingled wake-promoting and sleep-promoting neurones. Moreover, the sleep circuits are less well understood. One explanation being the rich connections in this brain region with the hypothalamus and in particular our old friend the ventrolateral preoptic area (VLPO). The VLPO has a great density of sleep-activating neurones, specifically it is a key part of the NREM sleep-promoting system. Xu and his colleagues go on to conclude that whilst GABA usually in the VLPO has an inhibitory effect on one arousal-promoting neurotransmitter; in the basal forebrain aspect it has inhibitory effects on all three of the wake-promoting hormones described. They conclude by stating: 'Thus, our results reveal a new pathway that promotes NREM sleep via broad inhibition of multiple wake-promoting cell types within the BF local circuit' (Xu et al., 2015).

Finally, we reach the cortex and focus primarily on the frontal lobe. The frontal lobe is, as has been discussed, exquisitely involved in sleep, dreaming, deep sleep, and in wakefulness. It is fundamental to who we think we are and our optimal functioning, it may be likened to the cortical soul. Less prosaically it is the cortical part responsible for the internal management of the mind and the external management of the person. It is frequently damaged in TBI and large bleeds, as well as certain strokes.

The pathophysiological basis of specific types of sleep disorder has only really begun being properly understood over the last 10 to 15 years. As described earlier much of the conducting of orchestras important for the initiation and maintenance of deep sleep, in particular, occur as a result of activity in the frontal lobes.

A few important studies stand as markers during this time period. Ioannides et al. (2009) were one of the first teams to describe these parameters in more depth using MEG data. They found regional activity especially that of gamma-band activity, was found in the left dorsomedial prefrontal cortex during deep sleep. Published in the same year Murphy et al. (2009) confirmed the importance of the anterior regions for slow-wave generation and distribution. They found distinct regions involved in their generation. As they state: ' ... slow waves preferentially involve several cortical areas including the inferior frontal gyrus, the medial frontal gyrus, the middle frontal gyrus, the insula, the anterior cingulate gyrus, the posterior cingulate gyrus, and the precuneus.'

TBI as an exemplar

The degree of importance of the anterior regions was underlined in a study by Koenigs et al. (2010) examined a group of 192 patients with focal brain lesions. Lesion analysis was decided through CT scans and neuropsychological testing and subsequent 'blind' reviewers. Lesion location data was then analyzed in relation to the severity of insomnia using an item on the Hamilton Anxiety Rating Scale (HAM-A). This was also a part of a later subgroup analysis that extracted those with greater mood and anxiety problems to test if these issues were confounding factors to the analysis conclusions. They were not; it appears that the damage to these areas in the brain due to the TBI was the primary (sole?) reason for the insomnia, discovered. Thus they found significant associations between insomnia and left dorsomedial prefrontal damage, that were not confounded through levels of mood or anxiety problems.

As demonstrated, the systems involved in the "battle" between sleep and wake and vice versa running from caudal to rostral and the reverse, are frequently disrupted in acquired brain injury and distinctly in traumatic brain injury. Of equal importance to TBI is its *sine qua non*, that of diffuse axonal injury. Indeed, some have suggested that the white matter damage is in fact more important than the direct coup and contra-coup associated with this type of brain injury. In that regard, even fewer studies have examined this kind of damage and its specific relationship to subsequent sleep disturbances. However, it is crucial to understanding it better. In addition, it helps demonstrate why in milder brain injuries; even in severe whiplash (at high speed) injuries, fatigue and sleep problems are so prominent.

Rotational forces at the time of impact cause a range of shearing injury to the cytoskeleton of the axon. Axons can be stretched, distorted, twisted, blunted, or actually torn if the mechanical force is high enough. After the impact the next 72 hours become one of the most significant periods for damage. This as described previously, is compounded at the neurochemical level, and not simply the result of the mechanical forces, as was originally thought. Certain zones of axonal bundling are especially vulnerable to this kind of damage. Dense bundling, as found in the deeper structures such as in the brainstem, is a prime example. As examined earlier, many orchestras, play over several networks, these rely on their functions for intact axonal pathways. This is especially true in the slow-wave sleep associated with later parts of N2 and the majority of N3 phases, or in other words the deep restorative phases of sleep.

It was to this end that Sanchez et al. (2019) looked specifically at white matter damage and how this damage was related to exactly that synchrony during sleep. They did this by examining patients who had suffered moderate to severe traumatic brain injuries. Within such a group various levels of white matter damage were found. This group was then contrasted with an age and gender-matched control sample of healthy volunteers.

In the first case standard protocols for assessing and then characterizing the degree of white matter damage were used. All slow waves were analyzed (peak-to-peak amplitude, oscillation frequency, and so forth). Finally, correlation statistics were applied to the data, with age as the covariate. The authors were surprised to find that the greater the white matter damage, the stronger the correlation with higher neuronal synchrony during N2 and N3 sleep, especially in the frontal and temporal regions. The same correlations were observed with increasing homeostatic sleep pressure. They go on to argue that higher white matter damage was associated with greater SWS power. This was also related in a direct way with fatigue levels. These problems were only discovered in the TBI patients and not the controls. Potentially, TBI increases synchrony during sleep. As an important aside SWS is always more synchronous; it is the degree of synchrony that is pertinent here. The implications of this will be discussed further in the next chapter.

Chapter 9

Chronic neuroinflammation and waste disposal

The long and the short of it

Further thoughts on neuroinflammation

Dwelling for a moment on synchrony. My first thought is the slightly tangential one, that epileptiform theories are replete with synchrony being one of the hallmarks of brain waves activity during seizures. These are so-called hypersynchronous states (Jiruska et al., 2013). Could this be a part of the explanation of why TBI, and to a lesser degree other acquired brain injuries, patients are at an increased risk of post-traumatic seizures? In other words, is synchrony more prominent during the parts or even the entire sleep-wake cycle, or is it simply 'easier' to get to; is the threshold lowered? What has been clearly shown is that NREM is the most synchronous of the stages of sleep. So the potential conclusion would be, people who have had a brain injury, are already more prone to greater synchronous activity, and are those with NREM problems the most likely to develop seizures post-brain injury? This synchrony was also previously linked to excitability and high carbohydrate diets in the original John's Hopkins literature (for a history see: Wheless, 2008). This stated that conversely low carbohydrate and high fats led to lowered synchrony, higher 'fluidity,' and a healthier brain. As we know in the early 20th Century, the only available effective treatment for epilepsy was the ketogenic diet. Unfortunately, it was time-consuming, hard to maintain, and at the time relatively expensive. Moreover, when compared to the lazy option of a daily pill; even with its concomitant serious side-effects, it lost favour. Interestingly, it has found a renaissance, apparently due to the side-effects, of the self-same medication, now being viewed as no longer acceptable (Elia et al., 2017; Hwang, et al., 2019). This is only half the picture, as what is less well understood is this and other diet's ability to reduce chronic inflammatory processes, and even more recently its ability to alter aspects of the circadian cycle. As discussed earlier, this is undoubtedly partly to do with the sub-component of Omega-3 fats: DHA, which increases hippocampal BDNF, which in turn leads to enhanced brain plasticity, membrane fluidity, and concomitant anti-ageing and anti-degenerative effects.

DOI: 10.4324/9780429199066-9

This naturally occurring enhancer of plasticity has been one of the few areas that standard medicine has examined. Research on so-called specialized pro-resolving mediators or SPM's. These are, in essence, derivatives of fish oil and PUFA's more generally. One such, arachidonic acid, has been the subject of numerous studies. The difficulty is that this competes with Omega-3 chemically, as an Omega-6 derivative. Our current Western diet is basically 30 to 1 in favour of Omega-6 (burgers, crisps, processed deli meats) over Omega-3 (e.g. fish, flaxseed, olive oil), and the human needs a 1:1 relationship. Thus far roughly the same anti-inflammatory effect can be had by taking a low dose aspirin, or better still, real whole oily fish three times a week as part of a good diet! (for a recent review, see Kraft et al., 2021). At the very least it is helpful that there is more of an interest in these chronic inflammatory processes, even if the results after over 30 years of research are somewhat meagre.

This also aligns with the work previously discussed on the so-called "primed" microglia, or heightened inflammatory status. If you recall, after a TBI (and other severe acquired brain injuries) neuro-inflammation is extreme during the acute phase, but is now viewed as lingering, far longer than was previously thought (Witcher, Eiferman and Godbout, 2015). This itself leads to the subsequent brain systems becoming more easily hyper-activated (through stress, illness, or indeed further acquired brain injuries) and thus producing secondary damage much more easily. The other thought is related to how frequently in clinical practice those of my patients using trackers (Fitbits and the like) demonstrate larger periods of deep sleep on their sleep data than their age-matched controls (so-called "benchmark" data). Thus, not only does TBI and associated white matter damage cause N2 and N3 sleep problems, it may even increase the potential for the duration of these stages of sleep; most especially the latter. The glymphatic system feels compelled to clean more and more, but is not able to fully clear away the debris during this extended period of SWS.

Chronic hyper-inflammatory states effect a number of specific as well as general brain and sleep functioning parameters. If you recall the work of Ramlackhansingh et al. (2011), chronic inflammatory processes (as measured through microglial indicators), have now been found some 17 years after a TBI. Thus in addition to increased inflammation through sleep problems, these come against a background of neuroinflammation due to the brain injury. This would potentially offer a better understanding of why there is a higher incidence of neurodegenerative diseases in the acquired brain injury population, due in part to the sub-group, who have both brain injuries and sleep problems. As stated, deep sleep deprivation leads to forms of cognitive impairment. Moreover, those who had difficulty maintaining sleep had an even greater chance of developing certain diseases of affluence. There are, as has been repeatedly demonstrated, many other implications for too little and too much sleep. As stated previously the protein *tau* associated with Alzheimer's disease is raised in those who have higher rates of sleep disturbance, most especially it is elevated in those patients who have the poorest NREM deep sleep. Although, as a brief aside, another theme in the book here, *tau* is actually helpful in the right amounts as it supports neuroplasticity. It is the idea of an excess that is problematic. Using spinal fluid assays which measured amyloid β kinetics, excess was found readily in sleep-deprived groups. They concluded that disrupted deep sleep increases Alzheimer's disease risk; not the other way around (Holth et al., 2019). It would be a useful study to assess the exact relationships involved in sleep deprivation and brain injury. No doubt, as shown, it is highly unlikely to be a direct one with severity. However, it would be interesting to tease out. In terms of clear direction, OSA has now been shown to be an independent risk factor for stroke. Chronic inflammatory processes have also been implicated, at least in part, in OSA. Outwith the more obvious chronic hypoxic damage in OSA, cycles of inflammatory events cause damage to numerous wide-ranging organ systems. Repeated hypoxic events and subsequent reperfusion injuries both activate inflammatory pathways. These then lead to irreversible vascular resistance problems. Very recent research has demonstrated a direct linear relationship with severity of OSA and the degree of calcified carotid arterial plaques (Chang et al., 2019). Chronic inflammatory processes are thus implicated in a number of sleep disorders, chronic neurodegenerative conditions, and acute acquired brain injuries. Most problematically in long-term follow-up these problems are found after acquired brain injuries.

As described previously, this information should, in addition, be tethered to the understanding of hypothalamic rest in REM during which it largely switches off. The hypothalamus is essential to so many processes but one fundamental one is temperature regulation. Inflammation of any kind produces higher temperatures. In addition, melatonin is, as Tarocco et al. (2019) argue in their review, a: 'master regulator of cell death and inflammation.' It is this relationship, which is one of the most important ones, that is highly affected after an acquired brain injury. As such further temperature, circadian rhythm, inflammation, and cell death regulation are going to be disturbed. One of the obvious effects of chronic inflammatory processes is that the hypothalamus has a harder time in regulating temperature. If you recall the work of Karaszewski et al. (2013), amongst others, after an ischaemic stroke higher temperatures in ischaemic and penumbral brain regions were associated with poorer outcomes. It takes little imagination to propose that as well as post acquired brain injury inflammation, related post acquired brain injury increases in temperature alongside the chronic inflammation would produce effects on sleep as well as other parameters. It would suggest that in addition to making it harder for the hypothalamus to do its temperature regulation, the question then becomes: when and how well is the hypothalamus able to rest and thus recharge in order to face the challenges of the following 24-hour cycle? This is especially the case as the hypothalamic areas were those with the highest temperatures, *irrespective* of the site of the initial ischaemia. Equally, it is pertinent to many neurodegenerative conditions, such as MS and Alzheimer's (Zindler and Zipp, 2006). It is also important to remember that this general elevation in inflammation and consequently higher temperatures were found in other chronic conditions such as ME, as well the TBI population studied by Jarred Younger and his team in Alabama (see for example: Younger, 2020). However, perhaps of even more importance is the work cited regarding individuals many months after the initial insult. Some 15 months after the initial stroke, when all would clinically appear fully recovered, or at least having reached a plateau, distinct microglia and macrophage infiltration were still apparent in all ischaemic tissue. The system damage had been done, it doesn't "know" when inflammatory processes should be reduced in intensity.

Whilst numerous efforts have been made to ameliorate these inflammatory processes in the early hours and days, few equivalent efforts have been made as these continue into the chronic phase. The examples previously cited from the work of Brait et al. (2012) are worth considering again. If you recall, blood-derived (as opposed to resident) macrophages are largely absent in the first few days after ischaemic injury but then gather pace to peak after the seventh day, after which they decline. Neutrophils on the other hand infiltrate the ischaemic tissue within the first 30 minutes then peak between one and three days. T lymphocytes whilst largely absent in the early minutes after damage has occurred, gradually increase in elevation between days three and seven. In other words, it is not simply a case of either inflammatory or anti-inflammatory processes, it is about the changing shape of these over minutes, hours, days, weeks, and months. What is also clear is the importance for recovery of some of these earlier inflammatory waves in the early days but not in the longer term. Indeed, I would argue, as we are only beginning to understand the various waves, as described earlier, of inflammatory factors, it may well be ultimately not only ineffective, it could cause harm. What I mean by this is that without some of these inflammatory factors, neuroplasticity and neurogenesis would become impaired. Without these, the forest fire, so to speak, new seeds will not germinate.

This is in opposition to my overriding concern in tackling later chronic inflammatory changes. For if this is not tackled, especially their causes, as they are the final common pathway in brain injury recovery, optimum neuro-rehabilitation gains will never be reached.

Deep sleep, spindles, memory, and brain injury

As has been shown other research has supported this view. In other words, there is a growing consensus that in addition to sleep changes being common for the vast majority of survivors especially the more severe the brain injury; NREM, deep sleep is prolonged. As was discussed earlier, spindles during NREM sleep serve a number of important functions, one of which is highly relevant to the TBI population; that of memory consolidation. As previously noted spindles usually occur on the descending aspect of each wave during sleep. They are widely held as important overtures to stage 2 and deeper NREM sleep. Although very brief, they are essential for memory con-solidation, sleep awareness and thus survival. In abbreviated form, NREM is hugely important for episodic memory consolidation and REM sleep is hugely important for procedural and emotional memory consolidation. Thus one of the hallmark problems post-TBI is memory for new events. If spindles are part of the crucial network involved in memory consolidation, it is unsurprising that, as they are damaged, new memories (and thus learning) will be harder to come by. In an excellent review, in 2012, Urakami de-scribed how spindles were a sensitive predictor of outcome from TBI. Overall, reactivity of spindles and basic activity of spindles measured in the acute stage were associated with better recovery. In other words, the greater the "noise" and frequency of "noise" the spindles were making during this period the better the final outcome from TBI (seemingly, largely irrespective of severity). Later work by Ducharme-Crevier et al. (2017), in the paediatric population seems to uphold these findings for children who suffered cardiac arrests and subsequent hypoxic brain injuries. In effect, the earlier the presence and overall activity of sleep spindles found in children recovering in ICU the more likely a good outcome occurred. Fernandez and Lüthi (2020) provide a review of spindles, both in terms of mechanisms of action and most significant functions. They conclude that spindles are not only im-portant for memory (in sleep and wake), they have wider significance for brain functions. The authors powerfully argue, spindles have so many functions, it is hard to reduce their importance to one primary one. They play a role in sleep, sensory processes, memory, neuronal plasticity, and other cognitive abilities. Most interestingly, their neuronal plasticity con-tributions outlast their EEG event. The thalamus and hippocampus are preferentially activated by them and these areas (along with the connections to the cortex) are critical for attention, and learning during waking hours. More strongly still they argue their own and other research cited by them suggest that spindles: ' ... fine spatiotemporal organization reflects NREMS as a physiological state coordinated over brain and body and may indicate, if not anticipate and ultimately differentiate, pathologies in sleep and neu-rodevelopmental, -degenerative, and -psychiatric conditions.'

Even if their arguments are only partially upheld in future research en-deavours, what is not in debate is the primary importance of spindles for memory consolidation. Moreover, good sleep spindle activity is not just a marker of an adequate night's sleep; it affects the waking cognitive activities of the person, most especially through waking attentional processes. This finding once again marks out the essential and intimate relationship between wake-sleep as a whole.

The other clear finding is that REM sleep, as a proportion of the overall length of sleep is reduced and is of poorer quality post-TBI. The latter has been evidenced through a decrease in eye movements (Busek and Faber, 2000; Baumann et al., 2007). Some studies have reported both increases and decreases, depending on which part of the night one examines in REM sleep but these are in a minority (Frieboes et al., 1999). What is less contentious is the conclusion that REM sleep, as a proportion of total sleep, is altered and appears (whether longer or shorter) to be of poorer quality after a TBI. Moreover, it is a commonplace that REM is disturbed in those with purely mental health problems, especially depression. It also points to the conundrum of what best to prescribe if these populations have the common co-morbidity of acquired brain injury and depression. As suggested by Wichniak et al. (2017) the clinician frequently feels compelled to prescribe both a Z drug alongside SSRI-based drugs to attempt (often unsuccessfully) antidepressant and an anti-insomnia effect.

Work by Ouellet and Morin (2006) and Parcell et al. (2008) supports the vulnerability of REM sleep to TBI pathology. They also point to a reduced REM sleep onset latency. This latter finding has been significantly correlated with mood disorders. Once again this has demonstrated too little or too much REM sleep has been associated with greater depression (as measured by scales like the Beck or Hamilton). However, much clearer and sustained evidence has been found for suicidality and shorter REM latency (see for example: Agargun and Cartwright, 2003). This has been supported by decades of research, to the point whereby it appears that greatly shorter periods of less than 90 minutes of REM latency (the interval between sleep onset and the start of REM) can predict suicidal ideation and attempts at self-harm in the following week fairly robustly (for a review, see: Palagini et al., 2013). As demonstrated by the same team emotional processing problems in a range of psychological disorders activates limbic system excitability in those with REM sleep disorders. Furthermore, those with both REM sleep deprivation and prolongation promote one or more type of emotional distress. In particular, the team's strong suggestion is that prolonged REM promotes a very specific type of depression, and this should be recognized as such diagnostically.

It points to the absolute fundamental importance of sleep again, and its vulnerabilities when brain injury strikes. As demonstrated, deep N3 sleep starts the process of cleaning and servicing, but it is REM sleep that completes this and it has to be the required amount, neither too much nor too little. Moreover, as sleep spindles can be viewed as the precursors to a number of later stages, it is very likely they play a significant role in the glymphatic system. Once again we come to a waste disposal problem.

More thoughts on waste disposal

The glymphatic system deserves some thoughts in a small section for itself. In future, this system will have much larger sections, even entire books devoted to it. For now the last five to seven years have witnessed an explosion of research in this area.

As I have demonstrated the brain is a very greedy organ, it consumes more glucose and oxygen than any other organ, as a consequence it generates a lot of heat and enormous quantities of waste products. Both CSF and blood have been hypothesized as the two main contributors to modifying this heat, which varies during the circadian cycle. The brain is generally higher in temperature during wakefulness and cooler during sleep. However, as I have also discussed certain critical areas of the brain are more prone to thermal sensitivity than others: the hypothalamus, the substantia nigra, and the hippocampus are three. Temperature dysregulation is both a product and cause of inflammation. Thus good CSF and blood flow are critical for cooling. Moreover, it has been hypothesized that the glymphatic system, along with its sibling the meningeal lymphatic system is dependent upon both adequate arterial pulse waves and the constant flow of CSF from the choroid plexus, as the two primary means of the influx. The glymphatic system runs in the same direction as blood flow, propelled by the blood flows waves. In other words, the beat of the rhythm of the flow of blood causes a mechanical pulse wave surrounding the blood vessel, to propel the cleaning fluids, obeying the laws of physical motion. Then CSF is mixed with interstitial fluids washing through the blood-brain barrier, which is assisted by the aquaporin (4AQP4) water channels. Eventually after washing away the toxins the fluid leaves the brain via the perivenous space and along the cranial and spinal nerves, becoming "traditional" in lymphatic clearance procedures. The process takes but a few hours. The meningeal lymphatic system may be described as the fast clearance system, in comparison. Drainage around the venous sinuses and the base of the skull can take only minutes to achieve (Mestre, Mori and Nedergaard, 2020).

Earlier I discussed the increased waste and toxins produced by any brain injury. Once again, the idea of so much being accommodated in a natural range, outside the "Goldilocks" points things go awry. Inordinate amounts of glutamate, beta-amyloid, tau, α-synuclein amongst others are produced. As well as this, every day we produce vast amounts of similar and other waste products that the brains glymphatic and meningeal systems have to clear away at night. The combination of the two waste product streams would overwhelm the system in any event. A good analogy which I use with patients describes the brain as having to deal with a giant toxic landfill site produced in a very short time by an enormous chemical factory: the acquired brain injury. In addition, this site is fed every minute of every day by a river full of further toxic waste: the simple process of waking brain function. At a certain point, the brain-injured person may have to simply deal with the river, in order to survive; leaving the landfill to destroy the local eco-environment around it. Or to move from the analogy to "allow" it to energize a perpetual cycle of inflammatory processes.

This would be all well and good if the brain injury did not harm in any meaningful way either of the two waste disposal systems. The latest research suggests this is far from the case. Christensen et al. (2020) demonstrated that whilst influx components are largely (but not wholly) unaffected, efflux components are frequently damaged within both of the waste disposal systems, but especially the glymphatic system, at least in TBI. From their own and others research (Bolte et al., 2020), it appears it is almost as though the brain signals for slightly greater influx, as it recognizes that there is a large amount of waste to be rid of. However, clearance of the same is much slower. More importantly for our threads of discussion, it is the limbic structures of the hypothalamus, hippocampus, and amygdala which are most prone to these errors in influx and especially efflux. In this sense, there is obviously a dual problem at large, one of which is simply a quantitative one: too much waste to deal with, the other one is the waste disposal unit itself has become faulty.

If this were not a big enough problem, the systems themselves are mainly clearing waste at night during sleep. This is especially true for the glymphatic system, which is in operation for 90% of the time during SWS (N3). Remember the studies on increased *tau* during sleep deprivation were most prominently found during SWS deprivation studies. The very thing that will promote healing in the brain-injured population, is actually a problem itself. Indeed, a large element of the foregoing and the potential connections made with temperature dysregulation, inflammation, and so forth may point to the glymphatic system being at the heart of these problems. Indeed, is the accumulation of waste mainly responsible for the accumulation of in-flammatory responses, at a certain point post-injury, this is intuitively very appealing. Tentative support for such notions can be found in animal models and research, with increasing concordance found in human research (e.g. Kylkilahti et al., 2021).

Repairing the glymphatic system and thus addressing the issue of waste disposal may help address the inflammatory problems.

Part IV

Treatments or promoting virtuous cycles

Chapter 10

Some ideas for research and intervention

Introduction

The aim of the first chapters was to focus on sleep, brain injury, and the combined effects of sleep alongside acquired brain injuries; in other words, to promote thought, ideas, and the author's own excitement at these respective fields and their potential, if these ideas were combined. The final chapter will hopefully offer something completely different (cue Monty Python music). That is to say the focus of the final chapter will be on potential practical translation of these ideas and to make proposals for their implementation alongside future research ideas. It is to this end that I will be deliberately provocative and passionate, and, for those that know me well, this should be unsurprising. Whilst the preceding chapters were intended to be a summation of the state of the art in research and theory, the final chapter will at times promote more speculative ways forward.

Assessment

People who know me, recognize that even though I am coming towards the end of my career, I still carry a number of soap boxes and torches regarding my chosen field of neuropsychology and neuroscience more widely. Indeed, these seem to have increased rather than decreased over the years! I hope that it is very obvious from the previous chapters that two of those most prominently are to do with sleep and acquired brain injuries and their relative lack of funding, in particular opportunities for patient recovery in the long term. In case the reader has missed some of these points, I would reiterate that most of the good quality research and understanding surrounding these issues has really only come about in the 20th century and most prominently there has been a significant growth in research only in the last 20 years, especially regarding the combined knowledge of these areas. As ever, what this means is that clinical practice lags somewhat behind

DOI: 10.4324/9780429199066-10

proper understanding of the issues at hand. I am proud to say that there are pockets of excellence in clinical practice across the United Kingdom, one of which is my own team (allowing for bias), whom since my first talk about sleep in 2013 has incorporated a better understanding of sleep issues within our population of acquired brain injury patients. I would go further and venture to say that each of the clinicians in the team from whatever professional stripe have a better understanding of sleep issues than the majority of clinicians I've come across in other settings working in neuro-rehabilitation. It has become standard practice to provide basic sleep management recommendations through our specialist fatigue management programmes, alongside a proper assessment of sleep issues from the very first meeting. It is here that I would make my first suggestion; that all those working in brain injury need to better encompass new ways of understanding basic sleep parameters, in addition to their standard practice of indicators of severity such as GCS scores and length of PTA. I would recommend that as part of the core data set of every neuro-rehabilitation unit or team, whether primary, secondary, or tertiary, clinicians should ask the following as a guide for better understanding and assistance to their patients. Basic questions should include this minimum data set:

QUESTION

Q1. Average time to bed to attempt sleep.
Q2. Average time to wake up from sleep.
Q3. Approximate length of sleep.
Q4. Any disturbances during the night.
Q5. If disturbed, are they at the beginning, middle, or end? Please tick all that apply.
Q6. Do you wake feeling refreshed?
Q7. Have you previously been diagnosed with a sleep disorder?

In addition to these specific questions around sleep, we have a separate section asking questions around fatigue issues subsequent to a patient's brain injury. This is particularly important, as I have shown during the course of this book, whilst some aspects of fatigue are undoubtedly related to sleep problems, others are completely independent of sleep and are sadly simply a consequence of the acquired brain injury that has occurred (e.g. bi-thalamic infarcts).

One impact of the aforementioned implementation is to start building a core data set that will then be able to be cross referenced with other indicators such as age, severity of ABI, etc. In time, even if it is done at a community level, will begin to enable us to have a better understanding of the prevalence, chronicity, and potential natural course with some linkage to indicators that are still not fully understood (e.g. the mixed data from severity of injury and kinds of sleep problems), which is sadly lacking at the present time. This statement is meant as an acknowledgement that although pockets of excellent research have been cited, large robust data from real-world databases is still needed.

As I have said previously, other than providing us with a more thoroughgoing understanding of the size and type of problem, it will also add weight to the need for a less crude understanding of what constitutes treatment options for those with such conditions. It is also the case that the majority of GPs; if presented with adequate information will more likely make a case to have a proper specialist assessment elsewhere. This brings us on to the concern of limited resources in a health service already stretched, which is a worldwide problem, not solely a UK one. To which I would say if you've understood any of this book, you would see that by resolving the difficulties around sleep and brain injury you would ultimately reduce the financial burden on the healthcare system, in the longer term. Significant chronic co-morbid problems start with perhaps only a single issue. At a very obvious level, both sleep and brain injury are dependent on the brain and its fully intact functioning. At the centre of this is the realization that we are fundamentally our brain. Dependent as we are on its ability to monitor, manage, and optimize every other function of our body. In time, especially with sleep and brain injury, they mature and develop into a host of other chronic health issues; which take up even more scarce resources. Thus, we end up treating, to name a few: further preventable cerebrovascular disorders, depression, hip fractures, cardiovascular disorders, type II diabetes, and a range of other distinct metabolic disorders. If we'd supported those better immediately after their brain injuries, with specific-tailored treatments around sleep as well as brain injury-related problems, some (in some cases, all) of these associated health problems may not have emerged or at the very least potentially stabilized their presentation.

Ignoring the problem will simply make it worse. It is with these thoughts in mind that I made a crude calculation of the number of specialist sleep centres around the United Kingdom. Irrespective of the source you calculate this from (e.g. narcolepsy UK, NHS England, etc.), there is a dearth. By this I mean it comes to 57 in one calculation through to a maximum of 129 on another. Given a population in the United Kingdom of approximately 67 million and the incidence figures previously cited in the book, this is woeful. What it is, is a disgrace for even if some acknowledgement were made of OSA and no other sleep disorder there is no way that this number of sleep clinics could cope with the number of referrals that could potentially be made in order to assist people with OSA treatments alone.

This of course brings us to the thorny question of diagnosis. For, if the dearth of specialist sleep clinics even at the upper end of my calculations is correct, there is not enough specialist assessments to make accurate diagnoses of specific kinds of sleep disorders in the general population, let alone those who have sleep problems in the brain-injured population. It is also true to say that the majority of sleep clinics are led by respiratory physicians, a number of whom I have met would welcome input from specialists in ABI. In this regard, I am unaware of *any* specialist sleep clinic for those with co-morbid sleep disorders and acquired brain injuries in the entire United Kingdom. We have thus a problem in terms of adequate assessments and proper diagnoses, within the general sleep-disordered population, quite apart from those who also have acquired brain injuries. In practice this means, as with myself and a few lone clinicians, we have to make the best of what we can in terms of our own understanding and diagnostic capabilities in this regard. So we end up, as with so many other difficulties, being in the position of advocates for our patients knowing full well there is something amiss and we are fairly clear on what that might be, but we do not have the ability to gain the rubber stamp or the required paper chitty or, in this day and age, a confirmatory electronic communication to confirm our diagnostic hypotheses that would enable our patients to be better helped. In this regard, I would make my second recommendation in order to foster this understanding and promotion of this area. I would suggest that at least one person or two people within a team be assigned as the unofficial or official sleep specialist(s). They should be familiar with standardized measurements undertaken by sleep clinics, in order to further support the acquired brain injury population. In addition to knowledge, they should also have use of standardized measures such as the Epworth Sleep Scale, the Stop-Bang scale, and other such paper and pencil methods of assessment, alongside which, and this really is in the real world, access to actigraphy measures that could be sent to patients of particular concern. All of these materials could then be combined into a very powerful, clinically meaningful, non-standardized, and standardized assessment of sleep disorders in those with acquired brain injuries that would potentially mean a greater change of onward referral to a specialist sleep clinic. Failing an onward referral, it may add the necessary weight to modifications to their existing treatment regimens.

A further (third) recommendation would be ABI Services to devote some time to training team members in basic fatigue assessment and management. As I've demonstrated, although there is a Venn diagram part where sleep and fatigue issues overlap, either side of this, they should be viewed as distinct and separate problems, which need separate approaches. Support for this idea is found throughout this book, extra support comes in the shape of the sweet spot, or 'Goldilocks' idea that I've also promoted throughout. Although akin to homeostatic responses and principles, it is better understood as disturbances of systems due to injury, that are not 'optimizing equilibrium' but rather a significant acute or chronic event or series of events that overwhelm the compensatory mechanisms and cannot be understood simply as a decompensation. Studies of various drug intervention trials have demonstrated distinct benefits in some aspects, but not all, depending on dose. Two good examples will suffice. A clinical example would be the well-known benefits of Mirtazapine at lower doses (15 or 30 mg) which more readily promotes sleep, but if anything, seems to also promote, or at least doesn't particularly help, daytime fatigue. A second would be the unusual effects of Modafinil. Rammohan and his colleagues (2002) found that lower doses at 200 mg per day of Modafinil reduced daytime fatigue and demonstrated sleep improvements measured by the ESS. However, when the dose was increased to 400 mg per day, only sleep and the scores on the ESS improved, whilst no improvements were shown on report or on measures of daytime fatigue. As with the 'fight' between the neurochemicals of sleep and wake, it is always a dynamic process. Too little will not be enough, too much may lead to overcompensation by the host, or indeed damage, but precisely what constitutes 'just enough' for each person is in greater need of precision than current practice affords. For much of my career, various discussions have been had regarding so called: 'tailored' drug treatments, but none have actually materialized beside the standard titration based upon the interplay between pharmacy recommendation and the 'average' host. Of related interest here is the continued suggestion of dopamine as an essential problem implicated in fatigue, after a wide range of brain pathologies (e.g. TBI, Parkinson's, and MS). This has been partly held up through dopamine agonist treatments in these disorders having repeatedly demonstrated some level of improvements in fatigue levels. The more problematic results of side effects and transient positive effects has meant these (e.g. Bromocriptine) agonists have not found themselves a clear and distinct usage profile in neuro-rehabilitation settings. Thus, whilst the dopamine hypothesis (especially as a result of basal ganglia damage) is widely held, especially for fatigue and motivation problems after acquired brain injury, the follow through has not happened for clinical practice. My own view, as discussed earlier, like the Schizophrenia hypothesis, should be reframed as to which dopamine circuit is largely responsible for these effects; a wide ranging dopamine agonist is, I would suggest, bound to cause problems at a certain level, rather like the old-fashioned chemical straightjackets of the 1950s in major psychosis. My other view is that the reason it may have unusual effects, other than the 'blanket' problem highlighted, is sleep is so complex that enhancing one neurotransmitter, more frequently associated with wakefulness, will only lead to problems with other neurotransmitters, especially those somnolents: 'in the fight' between sleep and wake. One only needs reminding of the counter-intuitive effects of serotonin and how glutamate can in at least one circumstance produce a GABA response that is actually excitatory.

The result of the proposed greater depth of training would be that the core of those in this field would have an understanding and experience of acquired brain injury, from this central foundation, extra basic training is already provided. This takes the form of the other most commonly met illnesses including but not limited to: diabetes (I and II), liver disease, substance misuse, cardiovascular disease, respiratory disease, and so forth. Now there should be in addition a good grounding in sleep, fatigue, and their particular presentation for those who have had an acquired brain injury.

Other than this pointing to the future, or onward referral, the more specific support that could be provided by a brain injury team or unit now follows.

Some drugs for insomnia – pills

Earlier discussion of sleep treatments noted that even when an effect was found for commonly used prescription medication, the side effects, long-term usage problems, and problems in rebound insomnia meant they probably shouldn't be prescribed in the first place. As important is the consideration of waking, indeed the wake part of the equation is more central to thought around treatments for those with an acquired brain injury.

Thus, the more common, and less effective in the long-term, approach to insomnia is medication. Although common, little robust evidence is available for most insomnia medication; except in the shortest possible usage time. The two most commonly prescribed drugs both effect the GABA alpha subunits. These are the classic benzodiazepines (BZD) and benzodiazepine receptor agonists (BzRA or the non-BZD). Whilst evidence for benzodiazepines has been found for the more modern short acting ones, no good evidence is available for long-term usage, indeed after only a matter of a few weeks increasing tolerance develops and addiction becomes commonplace. However, the real problems with all the benzodiazepines are the withdrawal effects. In his classic and widely cited paper, Pétursson (1994) noted that these effects seem to fall largely into three main groups. The first group can be classified as short-lived rebound anxiety, insomnia, and problems with concentration. These problems last between one and four days, but are heavily dependent on type and dose. The second group is perhaps the most common. It was described as the 'full-blown' withdrawal syndrome, usually lasting around 10–14 days, with a similar but more severe presentation to the first. Finally, there is the long-term group who can only fully recover from the benzodiazepine treatment and withdrawal effects through other forms of treatment (e.g. inpatient treatment programmes for addiction). This typology seems, in some respects, straightforward; however, it is worth noting the human cost of irresponsible prescriptions is thankfully far rarer today. Pétursson (1994) states:

Physiological dependence on benzodiazepines is accompanied by a withdrawal syndrome which is typically characterized by sleep disturbance, irritability, increased tension and anxiety, panic attacks, hand tremor, sweating, difficulty in concentration, dry retching and nausea, some weight loss, palpitations, headache, muscular pain and stiffness and a host of perceptual changes. Instances are also reported within the high-dosage category of more serious developments such as seizures and psychotic reactions.

Indeed, this description is almost a partial account from those available from recovered people who have been addicted to these pills. Some have described benzodiazepine withdrawal as the worst drug to recover from (e.g. American Addiction Centers). However, the benzodiazepines remain stubbornly one of the most frequently prescribed drugs around the world, especially in the United States and the United Kingdom. A more recent review, sadly over a quarter of a century later, largely agreed with Pétursson; Basińska-Szafrańska (2021) noted: ' ... even strongly motivated patients tolerate the process badly or experience early relapse'; this is even during and after formal specialist detoxification treatment, not simply self-managed or family physician (GP) overseen.

In their recent meta-analysis of reviews (64 good standard reviews of trials, and eight exceptional high standard review studies – utilizing the AMSTAR2 tool), Rios, et al. (2019) compared the effectiveness and safety of pharmacological and non-pharmacological interventions for insomnia. The paper concluded it was not possible to recommend any benzodiazepine, or any of those drugs that primarily effect the histaminergic (H1 specifically) system, for example, diphenhydramine. the many and varied side effects and negative primary effects. There were, in addition, very few treatments that could be recommended. They concluded that there was strong and consistently good evidence, from a wide range of multiple sources, trials and reviews, for only five forms of treatment. These five were: zolpidem, suvorexant, doxepin, melatonin, and cognitive behaviour therapy (CBT). Others, such as zopiclone and some short-acting benzodiazepines, were found, in one review, to be effective; however, the safety profile was less well supported. It is the latter problem that in my view effects these conclusions. I will not repeat the many and varied objections to benzodiazepines, and instead would briefly discuss the Z drugs. Suffice it to state the Z drugs are undoubtedly effective in the basic parameters of drugs used for insomnia, for example, sleep onset latency (SOL) timing. They have also been found to be less addictive than benzodiazepines, and cause far fewer (and severe) withdrawal effects. However, the emphasis should be on less and fewer. Indeed, both Z drugs have become controlled substances in many countries (starting with the United States) due, not only, to their increasing use as drugs of abuse, but also due to a lesser extent the concerns previously discussed of the increased morbidity and mortality risk. It is now illegal to possess them without a prescription in the United States. Zopiclones stereoisomer, eszopiclone, is, however, available and has largely similar effects, although its long-term safety profile is not yet known. We revert back to the idea that simply because a drug may suffice for a sleep study and basic RCT for insomnia, it does not necessarily mean it should be prescribed.

Similar concerns would be made against doxepin. It is an old-fashioned tricyclic drug, first patented in 1969. It has a standard score of 3 on the anticholinergic burden scale. Having said this, it has been demonstrated to have few problems of addiction and is generally held to have a good long-term safety profile in small- or low-dose regimes in the younger population. Overall, it is a better option than any other available tricyclic and has potential over other types of drugs that may be prescribed. However, it must be stressed this is only so in a low dose (i.e. usually 3–6 mg, in some cases, a maximum of 10 mg). Indeed, even with these low doses, those with acquired brain injury nearly always report increases in fatigue the next day; as with all tricyclics, it has a long half-life (approximately 17 hours). Thus, we come back to the more modern understanding of treatments that we should be aiming to rectify sleep–wake problems, as a whole, and just not treating one, either sleep or wake, in some strange isolation.

Suvorexant (Belsomra) has been discussed earlier but is worth briefly re-examining, as it is one of the new Orexin-based treatments that may prove more helpful. Unfortunately, numerous similar drugs have fallen by the wayside due to potential liver problems and unanticipated side effects on emotional regulation. In addition, it has a relatively long half-life. This could potentially mean a build-up of effects and side effects, it may be one reason for the FDA statement to Merck. The FDA in the United States ordered the makers, Merck, to reduce the dose as a consequence, from their recommended 40 mg down to a starting dose of 5 mg. This was found to be somewhat ineffective in the phase III studies. Even 10 mg showed useful improvements in objective measures, but was largely thought ineffective from subjective patient reports. There is another more important and related patient issue, this (and similar drugs) do not produce one of the main patient reported benefits of the Benzodiazepines and Z drugs, and the tricyclics, that is assistance in going off to sleep. In this sense, a significant proportion of those who have insomnia report difficulty falling asleep; something the Orexin-based treatments are very poor at helping. The main concern remains: the FDA field reports and later analyses that suggest, like zolpidem, it has the potential for abuse and seems to have specific effects on the emotional regulation system; especially an increase in suicidality (Jacobson, Callander and Hoyer, 2014). What is more, concerns have been raised through anecdote and mice studies that links of the emotional-cataplexy systems found in orexin-depleted humans can be activated through orexin antagonists (e.g. Mahoney, Mochizuki and Scammell, 2020). At best the most that can be said for these drugs is that we are a long way from understanding if they are useful for insomnia; caution is needed. Finally, all the drugs discussed earlier have effects on memory performance (encoding, storage, and retrieval) from the significant (anti-histamines), meaningful (benzodiazepines and Z drugs), through to the mild (orexin antagonists). As such they should not be used in the acquired brain injury population.

Melatonin has been examined previously and is probably the most easily supportable drug for insomnia. There is a high degree of consistency with my own examination of the literature and the recent meta-analysis of reviews by Rios, et al. (2019).

CBTi

Behavioural, non-pharmacological approaches have been successful over decades. This particular behavioural approach has a vast literature backing its power in over 50 years of research and clinical practice, across a spectrum of psychological problems. It should be done properly before all other approaches are tried, it is the foundation stone of good clinical sleep therapy. For an elegant meta-analytic paper, please see Irwin, Cole and Nicassio (2006), which looked at cognitive-behavioural, relaxation, and behavioural approaches to examine their respective differences and impact on the middle-aged and over 55 population, as previously identified at risk groups.

As I have demonstrated, the basis of cognitive behaviour therapy for insomnia (CBTi) is primarily behavioural principles (both classical and operant conditioned) properly applied. These latter two words are very important. As has happened with CBT approaches more widely, where once was an effective therapy offering hope to many, it has become a bowdlerized recipe book-driven approach, which, as it has become more well known to cure all psychological ills, is diluted or inappropriately used. A clear example is how industry is using it for enhancing work performance and reducing sick leave. A well-known bank, one of the largest in the United Kingdom and thus the globe, sends workers who experience the stress that their jobs give them for CBT to get them 'back on track' and reduce sickness absence. The reality is not due to their 'dysfunctional thinking' but the misery caused by their job. This is a clear example of how this therapy has lost its way. Sadly, mindfulness approaches are now becoming similarly at once a panacea and a problem if you can't do it properly. In other words, it is the individual that is blamed and clearly at fault, at no point a fault of a company, system, or a society. As noted earlier, psycho-education regarding sleep management and stimulus-control procedures have been demonstrated to be effective outwith any other interventions (Riemann, et al., 2017). These have been detailed in the first part of this chapter. At its best, over and above this CBT can assist with understanding your less helpful thoughts and behaviours that prevent you from sleeping well, specifically around sleep. It should also contain some tips, hints, and tools for stress management and relaxation.

Another arm of CBTi, and perhaps it could be styled as the third, or in Walker's (2017) estimation the most important, is sleep restriction or reduction procedures. A few words of explanation are needed at this point; in all respects, it is another separate component of a behavioural approach, and thus the two principal elements of CBTi are actually behavioural and potentially the most powerful components of the overall approach. In 2010 Spielman, et al., in their review of treatments for insomnia concluded it could be used as a stand-alone therapy, known as sleep restriction therapy (SRT), as it was so powerful. Numerous subsequent studies have further supported this view (see for example: Elliott, et al., 2014).

This idea should not be confused with the sleep restriction studies outlined in the first part of this book. They are studies designed to look at the impact upon various aspects of psychological and, more generally, physiological aspects of a particular period or periods of sleep restriction (e.g. on memory performance).

SRT actually does the reverse of the dictum prescribed earlier on regarding overall sleep opportunity time. What it is generally held to do is to work out the actual sleep time within the overall sleep opportunity time. For example, utilizing a sleep diary and more recently actigraphy, it may be demonstrated that the person only actually sleeps on average for four (4) hours a night. This then is used as the beginning of what is known as the: 'sleep window,' during which sleep and total time in bed are the same number of hours and minutes. Indeed, as with many sleep clinics, starting CPAP treatments, four (4) hours is considered the minimum for safe basic sleep. Approximately 30 minutes are added to this base time. So the starting point, in this case, would be 4 hours and 30 minutes each night. Over a number of weeks and months, this window is increased utilizing sleep efficiency scores to get the patient to increase their overall sleep time. Potentially this may be only a very small increase by perhaps only 15 minutes a night each week.

This is different to sleep compression therapy (SCT), which uses a different type of progression and systematic reduction in bed time to sleep time needs. Put simply, compression works in reverse. As over a number of weeks (frequently ten is chosen), bed time is compressed down to conform to the actual sleep time found in the first week of study. It doesn't always reach that point but usually stops at a level of balance between sleep time and efficiency. It has become more frequently used as a behavioural treatment over the last few years, as there has been an increase in adverse events and experiences with the older SRT method (bringing continuing daytime sleepiness as a major one) and SCT is considered a more gentle approach (as in the ground breaking study by Miller, et al., 2014).

It has been demonstrated to be effective (Maurer, et al., 2018). Indeed, time in bed regularity is now considered the most important aspect, not the restriction elements, leading one to think the gentler SCT as being the preferred option.

Finally, and most pertinently, what is the evidence for this treatment for those who have suffered an acquired brain injury? First, and most importantly, it brings us all the way back to the recommendations in earlier. This therapy should never be used with ABI patients who have been diagnosed with hypersomnia, sleep apnoea, or sleep conditions potentially other than insomnia. This may seem obvious, but sadly I have experience of a 'CBTi trained sleep expert' undertaking such treatment with a severely injured TBI patient. Once again, I come to the essential need for the expert to be dual qualified or trained and experienced in sleep problems and acquired brain injury. Looking at the literature on these treatments with appropriate ABI patients, some useful effects have been found.

A scoping review undertaken by Ludwig, Vaduvathiriyan and Siengsukon (2020) found five articles indicating that CBTi improved outcomes for sleep and some other co-morbid symptoms (e.g. anxiety), which finally recommended that a fully powered RCT was now warranted. Lowe, et al. (2020), in their systematic review of CBTi with stroke and TBI, concluded these therapies, along with acupuncture (with the vagus nerve as the target), were likely to be useful. However, a full meta-analysis was not able to be performed due to study heterogeneity. Finally, the most extensive and thoroughgoing of these systematic reviews was undertaken by Ford, et al. (2020). Of 4,341 studies found, only 16 were good enough to be included in the final analysis. Even here, they were cautious and did not like to offer firm conclusions, but did make it clear that CBTi did improve insomnia and sleep quality in the populations studied (predominantly stroke and TBI patients). Once again they recommended larger, properly powered RCT's to be instigated, as these and other reviews clearly demonstrated the potential for these therapies.

Herbal and over-the-counter remedies – some better pills and some water

Intuitively one may have thought that herbal remedies used for thousands of years may be safe and effective. Neither is necessarily true. Some herbal remedies are potentially unsafe and dangerous, even though they may be effective; perhaps the most well-known is St. John's Wort, which although very effective for mild to moderate depression can be dangerous in the wrong hands and dose. In severe depression, it can cause mania; in psychosis, it can produce florid symptoms. The theory behind this is that its effectiveness is largely based on its ability to increase serotonin in the brain (as with SSRIs). Having said this there is a degree of support for a number of herbal sleep remedies; especially for mild, moderate, and/or transient insomnia.

This is especially important, as I have shown, reliance on prescription medication for this is, apart from very few exceptions not recommended, in particular for prolonged usage.

The herbal remedies that have the most support will now be examined, beginning with Valerian.

Valerian has been recognized as a sedative and soother as long ago as ancient Greece, indeed Galen prescribed it specifically for insomnia. It has been noted as a herbal treatment in repeated herbal monographs including Culpeper's complete herbal of 1653. According to the food and drug administration of America (FDA), it is generally recognized as safe (GRAS). New research has demonstrated many potential candidates as to its soporific properties. Three in combination are considered most likely, these are: the plant derived form of GABA, valeric acid (from which older epilepsy treatments are derived), and isovaltrate. The first two are well known, the third is unusual and the more we learn about it, the more it would appear, it acts as an adenosine agonist and a serotonin modulator. All of these are in small amounts when compared to the isolated chemicals found in prescription versions. It is thus more to do with the particular mixture of these than their single influence. Further research is ongoing, and new compounds are being found to have sedating effects, for example, hesperidin and linarin contained within valerian. Indeed, when one examines the many chemical compounds that make up valerian, it is easy to see why they would in combination, at the correct dose cause some sedative effects. It is the latter issue that potentially causes us the main problem in evaluation. First, in reviews of reviews, meta-analyses, clinical trials, and cumulative evidential monographs, valerian is either found to have little or no effect or a direct effect that promotes various aspects of good quality sleep. When looking in detail at this literature, it becomes clear that dosage is a prominent concern in earlier studies. Most used sub-optimal doses of 300 to 450 mg at varying times prior to sleep. It is now considered that the starting dose for valerian to be effective is 600 mg; however, beyond 900 mg, there appears to be no material increase in benefits (Wheatley, 2005). In their extensive systematic meta-analysis published in the prestigious American Journal of Medicine, Bent, et al. (2006) concluded that valerian was effective at improving sleep quality and had no significant side effects at the treating doses needed. In a well-designed recent RCT study, Ahmadi, et al. (2017) found that valerian significantly improved sleep.

In a much more recent review by Mischoulon (2018), it was concluded that: 'In summary, valerian appears to be a promising hypnotic that decreases sleep latency and improves sleep quality, with potential niches in various patient populations.'

The National Institutes of Health (NIH) monograph office of dietary supplements (last update 2013), concluded that it is largely safe, although they did also state data on effectiveness was 'inconclusive.' Of course, the

NIH monograph lacks many of the important trials that have been cited in this book, in particular those over the last eight years. Finally, it is worth mentioning a well-designed triple blind study which points to the better kinds of study not available to earlier reviews of valerian. Published in 2021, Gholami, et al. demonstrated clear benefits of valerian in terms of sleep latency, duration, efficiency, and overall improved daytime function. Subjective and objective measures were used and allowed a firm recommendation to be made for valerian.

Overall, it is recommended that valerian be used at a dose of at least 600 mg (but no greater than 900 mg) for at least two weeks (before discontinuation) as a potential treatment for insomnia (especially for the delayed sleep onset type). Of even more benefit are valerian combination treatments. In this regard, valerian and hops has been the most widely studied. In a recent review, Yurcheshen, Seehus and Pigeon (2015) concluded that increasingly better quality research is demonstrating better effects therapeutically. In a full double-blind RCT, a mixture of valerian, hops, and passion flower was shown to be no different from 10 mg of Zolpidem over a two-week period. This is worth a pause, as Zolpidem, if you recall, was the most promising of all the Benzodiapzepines and non-benzodiazepine variants. In this no group differences between sleep latency, total sleep time, nocturnal awakenings, in addition to significant reductions in Insomnia Severity Index scores. These latter findings are especially important in the context of one of the main arguments of this book; that of the damaging effects of chronic inflammatory processes after an acquired brain injury. Each of the herbs listed for insomnia treatment: valerian, hops, and passion flower, are anti-inflammatories. They have been found to have effects on different and also related inflammatory systems, which would be another arm in our multi-layered intervention strategy post injury for improved sleep and better outcomes from the injury itself, and of course their intimate linkage.

A number of things should be borne in mind in these studies and reviews. First, there is no equivalent deleterious effects of these herbal mixtures when compared to the Z and non-Z drugs discussed earlier. Second, mixtures of herbal medicaments more readily conform to actual trained herbalist performance. It is more common that a herbalist will make a herbal medication based on their assessment of the person, which results in a number of herbal products combined. This is unlike homeopathy and standard medicine, where one single thing, chemical, or molecule is supposedly used to some effect. Importantly, the historical movement of valerian research (especially in combination) has moved largely from poorly designed (including ineffective dosing) through to average designs, to more recent good robust gold standard designs. As this has happened, support has moved from nil effect, through to possible effect and finally probable or clear effect. As with so much research, when the review or meta-analysis was sampled becomes critical. In this regard, it is also difficult to find specific research for herbal remedies in those who have suffered an acquired brain injury. It is here that particular combinations may be of more benefit. It is also essential that any such treatment is investigated properly to see if there are any interaction effects with the many (usually) drugs prescribed to those who have such brain injuries.

Finally, funding for good-quality research is sparse. Thus, findings of any kind are hard won, unlike the trillions of dollars spent on standard medical pharma. The total from the National Institutes of Health in the USA is around £125 million each year, and this budget covers everything from hypnosis through massage, tai chi, yoga, chiropractic manipulation, and herbal treatments. The latter sadly lumped together with some very odd and bizarre potential treatments. Overall, apart from recommending an ap-pointment with a properly trained herbalist, it seems sensible to suggest at least trying herbal sleep remedies (especially the combination of valerian, hops, and passion flower) for at least two weeks, as described. One final notation on quality, look for the THR label, since 2011 in the United Kingdom, this label ensures standards are the same as any other over the counter medicine. Of these, over 21 are based around the mixtures described and are specifically designed for reducing anxiety and/or improving sleep.

Sadly, herbal medicine is often lumped with homeopathy, as it is on the government website listing THR products; indeed in a shocking 2015 survey, two out of three people surveyed didn't know that herbal and homeopathic remedies were entirely different. In the author's view, the former is based on thousands of years of usage, needs considerable training to be prescribed (for specific patients), and has numerous good quality studies to back it up. The latter was developed by a bizarre individual, who believed many dis-eases were caused by coffee, and that basically water can cure anything from a cold, through depression to leprosy and most recently, in modern for-mulations: SARS-Coronavirus 2. As Sir Simon Stevens, the current CEO of the NHS summed up (2021):

> It's one thing for homeopaths to peddle useless but harmless potions, but they cross a dangerous line when making ridiculous assertions about protecting people from Covid infection.

If any reader was confused between these two very distinct methodologies, and treatments, I trust they are no longer. Herbalism works, it is the basis of modern standard medicine; homeopathy is: wibble.

There is variable evidence for single herbal treatments based on such things as chamomile, cherries, lavender, kava-kava, L-tryptophan, to name but a small number. Reviews have been largely less favourable, but again quality has improved and when in combination, some of these are likely to have better effects than in a simple single application. This book is not about recommending the entire herbal canon for various sleep disorders, rather that is a book in itself. What I would hope is that those desperate people suffering from sleep insomnia problems realize that there is hope from different treatments that need to be fully explored, ideally with a qualified herbal clinician, before any of the more damaging drugs are turned to that are available from large pharmaceutical companies via your local GP.

Drugs for hypersomnia and related conditions – some pills

As described earlier, a prominent feature alongside sleep apnoea and classical insomnia is hypersomnia. Indeed these three are the most common of the sleep disorders found in those who have suffered an acquired brain injury, representing some 82% of the total sleep disorders (Wickwire, et al., 2016). Crudely, hypersomnia may be viewed as excessive daytime sleepiness and/or sleeping for very long periods at night (a minimum of ten hours or more each night). If, through a careful assessment, qualitative disturbances (e.g. frequent need for micturition, sleep disordered breathing, and so on) have been ruled out, the clinician is likely to conclude that the hypersomnia is of central origin associated with the brain injury itself.

As I've established, no single, simple explanation exists for such central hypersomnia. Disturbances in the hypocretin and hypothalamic systems have been noted, as markers found in CSF. Most likely in my own view is an overactivation of the GABA and related somnolent neurotransmitter systems, which have become like this due to the initial surge in excitatory neurotransmitter systems caused by the initial insult (e.g. see earlier references to the glutamate systems). This would also make sense given the earlier discussion regarding potential decreases in histamine promoting excessive sleepiness. In some respects, an overcompensation occurs in the ordinary homeostatic functions of the competing sleep–wake systems. In other words, the beautiful conducting of orchestras becomes so disrupted, it never fully rectifies itself. There is partial support for such a notion, based primarily on the work of Rye, et al. (2012) and subsequent researchers, who discovered a subsection of patients who had suffered a TBI had increased markers for excess GABA in their CSF. These findings have led researchers (e.g. Lynne Trotti) to describe these patients as almost having chronic natural Valium or alcohol effects on their daily lives; obviously having significant consequences on cognitive vigilance amongst other important outcomes. As a consequence, a number of treatments designed to reduce the amount of extra GABA in the system have been tested. Trials of clarithromycin, flumazenil, and levothyroxine have all been tried with some limited success in this area, as they are, at least in the case of the first two, GABA agonists. The latter being a hormone has far more complicated outcomes. It has multiple effects, including the degradation of GABA, its release, and re-uptake, and importantly for the purposes of this and earlier discussion, the interplay with levels of glutamate. Of course, numerous side effects make clarithromycin and flumazenil problematic in other ways. The former frequently cause considerable gastric problems in the short term. However, in the long term, of greater concern are the effects on the gut biome. A recent review by Elvers, et al. (2020), found significant decreases in healthy bifidobacteria and certain species of lactobacillus, unfortunately these effects were found to last up to six months in some individuals. Flumazenil has the unfortunate common side effects of seizures, agitation, and headache. Given that both would have to be delivered repeatedly over the longer term to have consistent effects on hypersomnia, these are not issues that should be avoided in potential prescribing, and should not be glossed over.

Prior to this research, there were three primary pharmacological approaches to the problem of hypersomnia. One group has severe limitations, but the next two have some potential, along with controversies around usage. The first group are straightforward, powerful stimulants (e.g. amphetamine based drugs), the second group milder stimulants that promote wakefulness (e.g. modafinil, armodafinil), and the final group contain sodium oxybate (a standard treatment of narcolepsy). The first two groups largely work on the dopamine and noradrenaline systems, the latter seemingly (as it is still not fully understood) working on the GABA system.

The first group has shown some benefit, although these effects appear short lived. Moreover, most concern was raised regarding potential for amphetamine addiction, induced psychosis in treated groups, or at the very least relapse in previously stable psychotic illnesses. Indeed, in a recent review, the authors concluded that current discussion in the public and medical domain does not fully appreciate: ' … the extensive, and in many cases insidious, harms caused.' These harms are both physical (e.g. cardiovascular problems) and psychological (e.g. drug induced psychosis; Lappin, Darke and Farrell, 2017).

Turning to the next group, the knowledge of the use and effectiveness of this class of 'milder' stimulants is similar in history to valerian and melatonin. By that, I mean that concerns were raised about effectiveness, then raised about potential for harm, finally as better-designed trials and research has occurred, greater weight has been available for both effectiveness and safety. Certainly, in large measure, this may be to do with the impact and (relative to other medication) high-prescription levels for ADHD patients and its quite common use as a so-called 'designer cognitive enhancer' in the wealthy. All of this added to anecdotal power, which has now been translated into a more robust body of supportive research.

A number of more recent examples can now be looked at, Johannson, et al. (2020) is a good longitudinal study of methylphenidate use over five years in a TBI sample. They concluded that significant improvements were demonstrated in mental fatigue, depression, and anxiety for the treated group, whilst no significant change was found in the untreated group. The usage was found to be safe as well as effective. However, on withdrawal, significant deterioration occurred in exactly the same metrics, essentially meaning that this would become a drug for the remainder of the patient's life. These findings were virtually identical to a smaller study by Zhang and Wang in 2017.

A more nuanced and well-designed RCT using armodafinil undertaken by Menn, Yang and Lankford (2014) demonstrated clear benefits over the course of 12 weeks in the intervention group. Moreover, after this a follow-up using a 12-month open-label extension found these benefits continued. In addition, part of the design was clearly undertaken to find any adverse events, over both time frames in addition to any significant reasons for withdrawal. During the initial randomized phase, no significant effects were shown in either category, when compared to the control group. In relation to stopping, the two main reasons were for headache and anxiety. In the ongoing open-label component, those reporting headache tripled in incidence and thus reason for stopping. One person was admitted to A & E for a psychotic disorder, which stopped on withdrawal. Finally, those withdrawing due to increased anxiety were frequent. Thus, tentatively, it may be concluded that this study demonstrated very clear benefits in the use of the more potent form of modafinil for hypersomnia in the TBI population. However, it also confirmed that the reasons that people withdraw or have to have these medications stopped are consistent with earlier studies in the non-acquired brain injured population, that is, headache, increased anxiety, and potential for psychotic disorders developing. Thus, frequent monitoring is essential, which could easily be achieved through a local, community acquired brain injury service.

In a very recent review undertaken by Trotti and Arnulf (2020), they concluded that whether the hypersomnia was to do with an acquired brain injury or other of central origin, whilst the level of evidence for the treatment of such disorders was lower than in the more common classical narcolepsy syndrome, accumulating evidence supports the medications previously discussed. In particular, there is increasing support for modafinil/armodafinil. However, they also conclude that sodium oxybate is helpful as well. This is such a controversial drug both in terms of its addictive qualities and the significant issues around its use in date rape cases to make its ordinary use in clinical practice extremely problematic.

In addition, two new drugs have come to the market which also show particular promise. The first targets the classical dopamine noradrenergic systems, allowing both noradrenaline and dopamine to increase and is called solriamfetol. The other is called pitolisant and was developed by Schwartz and colleagues, the discoverers of the H3 histamine receptor system. It is, thus, the first H3 receptor inverse agonist to come to market and was only approved by the European Medicines Agency in 2016. Both of these two drugs have demonstrated in their clinical trials and in subsequent research to be powerful for reducing hypersomnolence, in the narcoleptic population. It would be hoped that trials of the acquired brain injury population will follow shortly. It should be noted, however, that solriamfetol can produce, in a significant minority of individuals, all the usual side effects noted for other stimulant medication i.e. in order of presentation: headaches, increased anxiety, nausea, and, in the case of solriamfetol, potential for the development of psychosis.

No such concerns come with pitolysant. In this regard, the H3 receptor inverse agonist is in my view a very exciting development. It is unlikely that it will produce psychosis, there is no reason to suppose given my earlier discussion of the specific issue around the dopamine hypothesis and psychosis that an H3 inverse agonist will have these serious side effects. These kind of conclusions are also supported by another recent review paper by Traeger et al. (2020) which specifically looked at the pharmacological treatment of neurobehavioural sequelae following TBI in the early course of recovery. Their conclusions were that whilst some support was found for the drugs highlighted previously (most especially modafinil), larger, better powered studies were still needed. In my own view, the evidence from a number of papers highlighted actually demonstrates that the level of confidence in these drugs should be increased now for specific hypersomnic problems of central origin. Indeed, I would make a plea for their wider usage as long as adequate monitoring was taking place to look out for the significant minority of patients who experience the adverse events discussed.

In conclusion, although the large majority of the treatments available for hypersomnia could be considered to consist of symptom mitigation and improvements in quality of life as a consequence of this, there are nonetheless valuable techniques that are currently underused for this population. The related, but more obviously amenable, causative hypersomnias such as those brought about through obstructive sleep apnoea and restless leg syndrome can more readily be treated from this perspective. An obvious example being restless legs whilst disturbing the course of a night's sleep. It has frequently been found to be due to lowered levels of ferritin and thus increasing iron supply can rectify this problem in over 30% of the patient group. Failing this, use of gabapentin has been demonstrated repeatedly to be of great benefit from a symptom mitigation perspective. In other words, for the primary problems that occur after a brain injury, there are treatments already available. Others should be utilized more given our understanding of usage over many decades and still others offer exciting hope for the future, most especially those targeting the H3 system.

Thus, to completely restore function or to ameliorate causative factors in something as complicated as ABI may not be possible. However, it seems reasonable given our current knowledge base to attempt to ameliorate hypersomnia's in this group starting with either modafinil or armodafinil and potentially even more excitingly, the new H3 inverse agonist described.

Hypersomnia – no pills

Whilst a number of behavioural interventions have been attempted, none has demonstrated significant improvements outside functional management of the hypersomniac problem. In this regard, they would at best be considered adjunctive support treatments more akin to accommodation strategies. These largely involve fatigue management programmes of one kind or another. Inevitably, they produce part of their therapeutic effect by simply assembling folk who have had similar experiences in a group. They do seem to show an ability to manage fatigue better, but do not impact upon the essential cause of the problem. As has been discussed, this is likely to be due to fatigue and hypersomnia being two distinct entities.

One cannot leave a section examining the effectiveness of treatments for hypersomnia without discussing the primary non-pharmacological treatment for this disorder, that is blue light therapy.

First, some words on what I mean: blue light therapy of one kind or another has been used to good effect for many years in a condition known as seasonal affective disorder (SAD). It would appear that a number of different neurotransmitter systems are affected by the application of such high intensity light (a minimum of 10,000 lux), in particular, the serotoninergic system. Measurable increases in serotonin have been noted in a dose response manner. In addition, more recent work has found that exposure to blue light increases the functional activation of the prefrontal cortex. This occurs after a minimum of thirty (30) minutes of exposure or longer. A recent study by Alkozei, Smith and Killgore (2016) found that even after exposure had ceased, persistent functional improvements in working memory were found through increased activation in the ventrolateral prefrontal cortex, as demonstrated on fMRI. In other blue light studies, increases in both serotonin and dopamine have been shown for decades, in addition to the more obvious melatonin and hypothalamic-pituitary-adrenal axis relationship with the SCN discussed earlier (see: Rao, et al., 1992, Oldham and Ciraulo, 2014, and then a novel approach by Volf, et al., 2020).

As described in an earlier part ABI (especially TBI), produces reduced hippocampal volume and lower levels of neuroplasticity and these can also be demonstrated in significant circadian rhythm disorders (Bedrosian and Nelson, 2017). Moreover, numerous studies over decades have supported the use of blue light for depression (even major depression and bi-polar disorder) in addition to seasonal affective disorder (Malhi, et al., 2009, Lieverse, et al., 2011, and Takeshima, et al., 2020). Indeed, blue light is now canonical in the psychological and psychiatric practice guidelines across the globe (e.g. Australian psychiatric practice guidelines 2020). What is more, it is now so mature as a treatment that RCT's have been conducted for some years against standard pharmacological treatments for SAD and standard depression (Sloane, Figueiro and Cohen, 2008). In a recent review and meta-analysis,

Geoffroy, et al. (2019), came to the strong recommendation that given the decades long support for light therapy (LT) as an effective treatment, it was odd that antidepressant drugs (AD) were still the first-line treatment. This was especially odd as well as LT being shown to be effective, they had a better safety profile and were cheaper in the short and long term for both seasonal and non-seasonal depression. Later in this paper they found that LT was non-significantly superior to AD, but that LT + AD when combined was significantly better than either alone. The final conclusion was: ' ... thus both LT monotherapy and combination [of LT + AD] may be proposed as a first line treatment in seasonal and non-seasonal depression.'

Sensibly, given this long history and specific understanding of LT's effects on particular neurotransmitters, largely the same neurotransmitters that are acted upon by the prescription treatments used for hypersomnia discussed earlier, researchers and clinicians began using LT for hypersomnia and related post acquired brain injury problems such as fatigue.

An early RCT was conducted in Australia by Sinclair, et al. (2014). This investigated the effect of four (4) weeks of the light therapy. The dose was 45 minutes of 465 nm (bright blue light), as compared with another groups yellow light (574 nm) and the control of no-treatment group. Interestingly, the conclusion from this study was that whilst a significant reduction in fatigue and daytime sleepiness occurred, no similar improvements were found for depression. The final recommendation was that blue light therapy: ' ... appears to be effective in alleviating fatigue and daytime sleepiness following TBI and may offer a non-invasive, safe, and non-pharmacological alternative to current treatments.'

It is within the study that we learn the reasons potentially why there are anomalies in various reviews and meta-analyses over the years. Contradictory findings seem to be found depending on the collection of studies examined (trite but true). Within this, the essential conclusion proposed by Bloom, et al. (2009) in their evidence-based practice guidelines is especially pertinent: 'Importantly, the most relevant dimensions of light therapy may be the timing, duration and intensity of the intervention.'

Earlier studies, which found little or no benefit, used limited time schedules (e.g. 15–20 minutes each day), limited time (e.g. two weeks total treatment phase), or very low levels of light (e.g. 3,500 lux). As standards have improved, more robust data has accumulated. Once again, much depends on the year of a review or meta-analysis. Standard administration now appears to be 30–45 minutes each day, of a minimum of 10,000 lux, for a minimum of four to six weeks. Moreover, given recent research from UCL regarding standard SSRI time frames for effects beginning after around 6 weeks for anxiety but not depression till around week 12; it would seem that something equivalent would be needed to understand properly the effects of LT. Indeed, numerous modern studies and reviews essentially make the same points. As Onega and Pierce (2020) conclude in their review (which includes practice suggestions): 'Bright light exposure is a safe, non-pharmacological treatment that is currently underutilized in this population. Clinicians may find bright light therapy beneficial as a primary or adjunctive treatment in reducing depression and agitation in older adults with dementia.' This is also true for TBI and now Parkinson's disease (see for example: Raikes, et al., 2020; Endo, et al., 2020). It seems that a wide range of non-progressive brain pathologies, can be at least functionally ameliorated by the proper application of the right kind of blue light, at the correct intensity for a minimum of 4–6 weeks.

Returning to the central theme in this section of hypersomnia and the related problem of daytime fatigue, Anders West and his team in Copenhagen have explored these specific interventions for some years. A fine RCT (West, et al., 2019) is a good exemplar. The study starts from the premise that neuro-rehabilitation centres, and thus patients, lack exposure to natural light. Post-stroke patients were assigned to a purposely designed intervention unit with naturalistic lighting and to a standard control rehabilitation unit. In the intervention unit, the patients demonstrated significant, clinically meaningful differences in fatigue, when compared to the standard unit. In this recent paper, the authors were concerned that given the knowledge we have from sleep studies and years of studies regarding other pathologies (e.g. depression), only six studies could be found which specifically looked at light (mostly blue light) in a hospital setting. Three had neutral findings and three were positive when blue lighting was artificially altered to become either more naturalistic or in bursts highlighted earlier. They also noted the more bizarre, in this context, better-known assertion that too much exposure to the wrong kind of light, or even the right kind of light (blue) at the wrong time (during the night) will cause various harms; one well-known one being greater fatigue (e.g. Bernhofer, et al., 2014; Cho, et al., 2013). An even more powerful study was again clearly demonstrating this negative effect of the wrong timing of the right light. The study by Cope, et al. (2021) found elevations in acetylcholine and reduced dopamine expression in mice targeted with blue light during their sleep period. These elevations and reductions were then found to increase punishment perseveration and forced swim test immobility. Before the spectre of the poor translation between mice and human studies is raised, these behaviours and associated neuro-chemicals are exactly the same ones used to test and measure most of the SSRI's currently used in practice. The study and its forebears are not easily dismissed.

That some authors are still debating the basic premise outlined seems almost perverse. We should now be at a stage whereupon we are actually refining the targeted blue light therapy approach for what problem at which particular time. The year2020 was a very promising year for bright light therapy research and people recovering from brain injury. Two good examples exist of treatment with blue light, which will now be unpacked for detail. Salva, et al. (2020) undertook a RCT using blue enriched white light therapy specifically for reducing fatigue and hypersomnia in survivors of severe traumatic brain injury. Standardized measures included the ESS, the Fatigue Severity Scale, and the Pittsburgh Sleep Quality Index. The new standard of 30 minutes, of a 10,000 lux lamp, over the course of four (4) weeks was used to discern any differences between the intervention and control groups. The conclusion was an unequivocal support for the use of blue light therapy as it reduces fatigue, improves sleep quality, and daytime sleepiness. At the opposite end of the TBI spectrum, in a paper already alluded to, Raikes, et al. (2020) delivered the same dosage (10,000 lux) for the same time (30 minutes) for six (6) weeks in a placebo-controlled randomized trial to those in the intervention group who had sustained a mild traumatic brain injury. The main outcome measures were Epworth Sleepiness Scale, Pittsburgh Sleep Quality Index, Beck Depression Inventory II, Rivermead Post-concussion Symptom Questionnaire, Functional Outcomes of Sleep Questionnaire, and actigraphy-derived sleep measures. Large differences were noted in improved hypersomnia. The authors concluded that: 'These findings further substantiate blue light therapy as a promising non-pharmacological approach to improve these sleep-related complaints with the added benefit of improved post-concussion symptoms and depression severity.'

In my own experience, it appears that for a largely unpredictable subset (at present) of patients suffering from hypersomnia post injury, blue light therapy, at exactly the standard time, lux, and duration set out by Raikes, et al. (2020), can work very well. Indeed, one patient has repeatedly told me that in his view I 'saved his life' because of this treatment. For just as many others, it appears to be a 'waste of my f***ing time' to quote one of my very frontal patients.

I am currently trying to work out who will benefit from this, prior to formalizing a research project later in the year. At the moment all, I would say is that it doesn't appear to be related to injury severity, but I am happy to be proved wrong on this specific issue. Of greater importance is it undoubtedly works very well for some. In this regard, it would appear to be a more sensible starting point as a treatment option, than leaping immediately to prescribed medication. For although, as I demonstrated earlier, unlike for insomnia, prescribed medication *is* an effective approach for hypersomnia, it has potential for numerous unwanted side-effects.

Finally a word on natural sunshine. It is better, in most ways, than a light box. Indeed, as I'll demonstrate later in a highly important way. It also has, on a sunny day, many more tens of thousands of lux available to the person outside in it. However, for the person with a TBI, inappropriate instructions for 'natural cures' can backfire. In this regard, one of my patients found themselves in the hands of a 'sleep expert psychologist.' Unfortunately, this individual dismissed the light box he had been using to some very small benefit in the pursuit of standard CBTi, with all the usual treatments and an instruction to get outside for at least 30 minutes of natural sunlight each morning; no matter the weather. He dutifully walked for the required time each day and felt more and more tired, increasing both the immediate and 'downstream' fatigue, so often found with those who have suffered a severe TBI. Once again, I caution from the general principle to the specific patient, things frequently go awry. In this sense, a very small benefit to hypersomnia was replaced by a very large cost. In another example, a different patient has carefully used his light box to gear himself up, especially in the late autumn, winter, and early spring. Fading in natural light later in the morning to reap the maximum benefit for both. Although he likes walking, he takes the car out, walks for perhaps five to ten minutes and then rests. The whole 'light' treatment operation takes up most of his morning, but has enabled him to feel human once more. As he said, the focus may seem long-winded for some, but given how wretched and unable to do anything at all, he felt before, it is a 'miracle.'

In summary, hypersomnia after any brain injury is all too common. According to a recent Headway (2019) summary of the academic research and their own members, it (or fatigue) affects over 70% of those who have suffered any type of acquired brain injury. The good news is there are a range of potential treatments, most of which are seriously underutilized in this population. In contrast to the overused prescription medication used for insomnia, which is poor, the correct prescription medication in hypersomnia is effective. Moreover, in both types of the most common sleep problems post brain injury, there are underutilized treatments (starting with melatonin and then herbal methods for insomnia) and non-pharmacological treatments (starting with blue light therapy for hypersomnia) that should be used before prescriptions are issued. Irrespective of these specific treatment options, there are foundational principles that should be followed, some better known than others to which I will now turn. In other words, I trust that I have outlined the three primary treatment options for the three primary sleep problems found after any brain injury. Those are insomnia, hypersomnia, and sleep apnoea. Beneath these, there is more to basic sleep management than may be supposed, especially if we are also going to target the related problem of chronic inflammation.

Chapter 11

First foundation

Introduction

The basis of any sleep programme or proper treatment should adhere to the fundamentals of sleep management (hygiene), Vitamin D3 levels, improving the gut biome, vagal nerve stimulation, and related strands emanating from these four foundations. In other words, what follows are the fundamentals that should be adhered to in any treatment plan, irrespective of the kind of sleep problem being treated. They follow the same principles, that brain injury and sleep should be treated together, from the same stance.

Sleep hygiene is the rather silly and certainly misleading term for the simple, yet highly effective idea of increasing sensible routines and good habits, and then the wider idea of further related environmental modifications. From now on the reader should note that I will be using the term sleep *management* instead, as it is what it says and cannot be conflated with sanitation!

It is very important that sleep management programmes, instructions and so forth should not be thought either as a panacea or the same for everyone. From the preceding, and I will repeat this with examples, general principles should be tailored very much to the individual and their specific needs. It is important to test, measure, think, recognize the need for seasonal change, and above all perfect the ideal programme. This process can be sped up through working with a sleep expert, but as I shall demonstrate later, this expert should have wide and deep experience of acquired brain injury. Standard sleep expertise will not suffice. It is incredibly important to, at the very minimum, go for an assessment by a dual qualified sleep and acquired brain injury expert, as they are more likely to identify a specific sleep disorder, that no amount of management will fix (e.g. sleep apnoea). The problem by not going to do such an assessment will at best cause prolonged frustration at a lack of progress, at worst it may cost a life.

Prior to outlining the absolute fundamentals, I shall note important and well known considerations.

DOI: 10.4324/9780429199066-11

Food, drink, and smoking

Avoid large meals; some have suggested a small carbohydrate snack is acceptable, it is not, and it can lead to larger snacks. It is best to simply avoid all food, as a minimum, one hour before bedtime, this harmonizes with a later suggestion regarding blue light devices. However, most importantly avoid large meals at least four (4) hours before bedtime, especially heavy, fatty or very spicy ones. Caffeine can happen after your lunchtime meal, but no later and certainly not with an evening meal. Caffeine, throughout the day has been shown in some limited way to be helpful for those with hypersomnia, so once again, the instruction is be guided by a clinician who knows the basic diagnoses. Alcohol should be avoided before bedtime, due to the rebound effect, which disrupts sleep later at night: follow these instructions. If you are one of those rare people who still smoke (stop it!), then at least don't smoke an hour before bedtime, nicotine causes sleep disturbance, but has a relatively short half-life (two hours). Avoid any fluids (including water) one hour before bedtime, again this harmonizes with other recommendations of not doing certain things one hour before bedtime. Overall, the one hour before bedtime should be viewed as the wind down, relaxing pre-bedtime routine, before the full routine happens.

True sleep opportunity time masked by the time spent in bed

Calculating a proper sleep wake schedule, which includes taking off time to allow for standard periods of wakefulness, is not always undertaken properly. This means due allowance for the essential sleep opportunity time minus the average total wake time for an individual based on their age. As I've shown, everyone has wakeful periods, most being a matter of seconds, some up to two–three minutes. What is equally crucial is that normally, none of these are powerful enough to disrupt us completely from the stage we are in (except when very old or very young), but when summed they amount to a considerable period of time. As I have highlighted previously this is a simple but crucial calculation, all too frequently overlooked. This will be different for different ages. However, reliable data is hard to come by. In addition, to the academic work of Arnardóttir, Þorsteinsson, and Karlsson (2010) and later Klerman, et al. (2013), the most reliable data we have available is from the so-called 'big data' sets of sleep-tracker companies (e.g. Fitbit). Waking and wakeful periods and the overall total are a much under-examined area of sleep and in my view may be helpful in explaining a number of phenomena. This is especially odd, as long ago as 1970, one of the early pioneers of sleep studies Kleitman produced a paper entitled: 'Study wakefulness. Study the rest–activity cycle. Don't just study sleep.' In my own view, this should be interpreted primarily as wake in the daytime, but it should also mean wakeful periods at night. Sleep–Wake is all of a piece. From what we do know, awakenings in total are more extensive and frequent than perhaps we have previously realized. Thus overall, according to the sleep-tracking data, adults spend cumulatively on average 55 minutes, awake at night. From the few academic research papers in this field it is clear that this matches up with smaller samples of polysomnographic data. The latter found similar averages, but also that awake frequency and totals conform to a U-shaped curve when plotted for age. What this largely means is that the youngest and oldest studied have the most frequent and highest totals. Thus someone who is around 13 years of age will likely have the least frequent and total wakefulness period(s) whilst asleep. The phrase often heard from parents of their adolescent offspring is indeed a truth: 'A bomb might go off and he wouldn't wake up!' Their grandparent, however, may be awake for many more significant periods and the total may accrue to one and a half hours, even to two hours. However, if we use the average, we need to take off around one hour from the bedroom total, as a crude illustration. Thus, the patient may say they go to bed at 11.30 pm and then get up at 6.30 am and express the view that they don't understand why they are 'always' tired. Taking off this hour means simply they have actually achieved only approximately six and a half (6½) hours of sleep, and as we know the aim should be for the band between seven (7) and nine (9) hours. They are thus sleep deprived purely through not calculating the proper total of their sleep opportunity time, including a good average for being awake during the night. True sleep opportunity is not simply based on bedtime and wake time.

The next thing to do properly is to figure out, as best one, can if the person is more of an owl or a lark, pre and post-brain injury, these may have altered. Then further discussion can be had around bedtime and wake time. In this regard getting an owl to potentially go to bed at a lark's time of 9 pm will risk halting any progress you may later make with medication, or non-pharmacological treatments specific to that individual. These foundations are exactly that – to be built upon.

Behavioural routines shape neurotransmitter cascades

The next thing is to decide on a time for going to bed (we'll think about actual sleep in a moment) and after that the time for waking up and getting out of bed. The latter should not be debated: brain injured or otherwise, after the alarm has sounded get out of bed and get upright and go to a different room for fully waking up. The reasons for this very clear, even harsh, instruction is based on a number of studies and also behavioural shaping procedures. First, if one hangs around pressing the snooze button, or lolly-gagging in the bed half awake and half asleep, the brain gets confused and thus its neurotransmitters are unsure if they should fully fire up the excitatory ones or keep the somnolent ones going. Second, and relatedly, the bedroom is for sleeping and sex – nothing else. In getting up and allowing the excitatory neurotransmitters to come fully into effect, after the RAS has become upright (or at least higher than 40 degrees), the behaviour will over time automatically enhance the appropriate excitatory neurotransmitters on the command of the alarm. Further praise to Pavlov, classical behavioural conditioning and his salivating dogs to the sound of bells at this point.

The human personality, especially that of the more modern one, may crave novelty, excitement and change, but the body craves the reverse for sleep time and wake time. The human body, in contradistinction to the modern personality, at least for the purposes of optimal sleep, benefits from a cast iron bed time and wake time. Consistency is thus key. Go to bed at roughly the same time each night (within a 30 minute window), and set the alarm to go off roughly at the same time each morning (again within a 30 minute window). Having said this there will be occasions when, perhaps due to an exciting day, sleepiness is outside these strict parameters. So as long as this is a one off in the context of the rigidity usually apparent, then bedtime can be stretched to accommodate a time when genuine sleepiness occurs – very occasionally.

Routine is more than simply a clear bed time and wake time. For maximum benefit, the routine should be anchored by sub-routines. These will foster greater propensity for actual sleep. By this I mean furthering the classical and operant conditioning. As we know from either side of the behavioural spectrum, clear physiological, or indeed in this case, neurophysiological connections will eventually firmly result. At its simplest this could be seen as behavioural chaining i.e. one thing following another, each cementing the other through repetition. A clear four stage common process will provide an exemplar. Prior to bed, a person may reliably relax in their sitting room, listening to gentle music or the radio (no blue light television for the hour prior to bed time – the first winding down phase), clean their teeth, then undress and put their nightwear on, following this they listen to the Radio 4 Shipping Forecast, then to sleep. These actions can then be broken down to each component part. Indeed, behavioural chaining, as used in neuro-rehabilitation settings relies upon the breakdown, and the assembly of numerous such routines.

A good example of this applied behavioural analysis (ABA) approach would be for cleaning teeth. Behavioural chains are sequences of discrete small behaviours that when linked together form a terminal or complete behavioural repertoire. When teaching behavioural repertoire using the method of chaining the initial part always involves the thorough completion of what is called a task analysis. Task analyses primarily serve to identify all the discreet small subunits or shall we say teachable units that go together to make the whole behavioural chain. The task analysis for brushing teeth may depending on the level of each subunit comprise anywhere between 15 and

25 subunits. For example: 1. Take tooth brush out of cabinet, 2. Take toothpaste out of cabinet, 3. Remove cap from toothpaste, 4. Hold toothbrush with left hand 5. Grab toothpaste with the opposing hand, 6. Place a pea sized amount onto the bristles of the brush, 7. Turn on water and so on and so on along the behavioural chain.

If done roughly in the same sequence, at the same time, each day these subunits are; to once again use the earlier phrase, fired often enough will become wired. It is important to note, that once wired if these actions are repeated the wiring process will bring about repetitive forms of neurotransmitter cascades. These cascades enhance the wiring and the potential for the next anticipatory sequence in a chain. Coming back to bedtime routine, if cleaning teeth is one of the larger behavioural chain subunits in the four-stage process previously described then the end subunit and last part of the chain of cleaning teeth will extend the example, become intimately linked and thus eventually wired with undressing and putting on nightwear and so on and so on until the full realization of the whole night time behavioural repertoire is complete. The final part of this routine, for some people may involve relaxation techniques (e.g. the 4–7-8 breathing technique, or any of those described in the review by Varvogli and Darviri, 2011). Indeed, it could even involve the new panacea of the modern age: mindfulness exercises, or some other slowing down technique. Whilst there is a wealth of evidence for behavioural routines and sleep enhancement both in the general population and for those with an acquired brain injury, it is only within such routines that relaxation therapy or training appear as important. Most recent reviews and studies are generally neutral or equivocal (Minen, Jinich and Ellett, 2018; Miller, et al., 2019). Indeed, when weighing the research properly, there is marginally better evidence for the essential oil lavender as a sleep enhancer; an improver of both quantity and quality (Fismer and Pilkington, 2012; Lari, et al., 2020). However, folk absolutely swear by various relaxation techniques efficacy. My own reading of the literature and indeed the earlier literature on relaxation therapy *per se* is, at its heart, are the breathing techniques properly applied. In that sense any fully trained breathing technique will suffice, especially if tethered to a good routine. Hence my earlier recommendation of the 4–7-8 technique. It is simple to learn and highly effective, largely because it is a rehashed modern version of the ancient method known as pranayama; which I'm not referencing, as it is approximately 7,000 years old. It has been touted as to the 'best cure' for all sleep problems, it is not. It is however, easy to learn and probably more effective than the equally popular box breathing technique.

I am rather against bedtime reading, music or speech from a radio, as a final component, on the grounds of purity of behavioural shaping. However, if it means the difference between staring at the blue screen of a mobile phone, laptop, or indeed a head full of anxiety promoting thoughts, then any of these three are lesser problems, simply better achieved through dimmed non-LED lighting, in the case of reading, or ensuring the radio is switched off prior to sleep in the case of the other two. However, I would still have the goal of no reading or radio in the bedroom if at all possible (remember sleep and sex).

Routine structure and behavioural conditioning also apply during the night and in the morning. If every time, which as you now know is quite frequent, you wake up in the middle of the night but rather than dismiss this (if one can), one instead starts to routinely fret and worry, stay in bed tossing and turning, allowing the anxious or other thoughts to grip and tighten their hold; a new unhealthy set of behavioural shaping may begin. In this sense it is a difficult balance between two courses of action. Stay in bed for a few minutes, to not worry overly and go back to sleep. Or forget staying in bed and tossing and turning for too long, and going and making a warm milky drink. After approximately 20–30 minutes, it may be better to get up and change mental gears. However, this should not be a hard and fast rule to shape behaviour. The reason for this is simply that if this occurs too frequently, the expectation will actually occur subconsciously and make the individual rise in order to get the milky drink. Just as good habits can be frequently fired enough to become wired, faulty and dysfunctional habits can be trained in similar ways. This can be easily understood in a more common way: it is remarkable the number of people who report waking up a few seconds just before their alarm goes off, if they have got a regularity of bedtime and wake time habits. To conclude, as with the idea of sleep opportunity time that I have previously discussed, routine and structure seem on the surface an easy fix. If sleep management is done properly it is not easy, it should be forensically examined. After such an examination, a properly built environment (internal and external) along with a complete set of behavioural routines will become second nature, require virtually no extra effort, yet provide enormous sleep, immune, anti-inflammatory, cognitive, mood and overall health benefits for those recovering from an acquired brain injury. Having said this be gentle. No one will be able to make too many changes at once that will work. The body, including the brain, does not like rapid metabolic change and will 'fight' it, the best changes come through gradual habit and routine changes, which eventually become permanent. The next key aspect of a thoroughgoing sleep management approach is an examination and potential manipulation of the environment in a house and more specifically, in a bedroom. The three primary factors that interfere with a good night's sleep are three of the primary factors involved in sleep, full stop. These are temperature, noise and light.

Temperature inside and out

I made brief reference to the temperature in a house and specifically in a bedroom earlier. As we know reduced temperature internally is one of the ways that the body, (through the action of the hypothalamic pathway), instigates the process of sleeping. Crudely, it has been demonstrated that whilst there is no full consensus, most studies found the optimal temperature seems to range between 15 and 19 degrees Celsius for the most comfortable sleep. This allows for the natural dip in core temperature as the evening progresses but crucially if the temperature is set to be lowered as the evening progresses, especially from eight to nine o'clock onwards, it is another way that environmental cues can on the one hand, harmonize with the body's natural circadian rhythm, but as with other zeitgebers, help the entrainment through further behavioural shaping. The difficulty here is to understand your own temperature profile, by this I mean again, if sleep management programmes are done well, they do not simply rely on generalizations around temperature. The best example of differences in optimal sleeping and temperature of the bedroom environment are based around differences in age. A very young person (infant) or very old person would need a higher environmental temperature. At either extreme of age, temperature regulation is more of a problem. However, in older age another related problem also tends to occur and that is best described as a cooler running temperature internally. Thus, a fully fit, athletic, 30-year-old person would more readily benefit from a temperature at the bottom end of this range i.e. around 15 degrees. Whereas a more sedentary retired person in their seventies would most likely benefit from a night-time temperature of 19 degrees.

Moreover, these age ranges are further complicated by health conditions which may mean that the individual is, or at least feels, more comfortable with a higher temperature. This poses a practical difficulty as on the one hand they may perceive this to be of benefit and indeed the heat may well be of actual benefit (e.g. in some chronic pain syndromes) to them in the day, but in all likelihood if too hot at night, the reverse will be true and their experience of a disrupted sleep will be more likely. What is less well understood is how higher temperatures specifically affect different qualitative aspects of the sleep experience. A good example is how higher ambient bedroom heat can affect slow wave sleep. In a series of studies by Kräuchi and his colleagues at the universities of Basel and Turin, they have demonstrated clear improvements in SWS, if support is given environmentally (through temperature alteration and different mattress types). In their most recent study, an RCT published in 2018, Kräuchi, et al., finally stated:

> 'In conclusion, the study expands the previous findings that a steeper nocturnal decline in CBT [core body temperature] increases SWS and subjective sleep quality, whereas inner conductive heat transfer could be identified as the crucial thermophysiological variable, and not CBT.'

Further experimental support for these ideas have come from Ichiba, et al. (2020). They used warming and sham eye masks before sleep to modify physiological heat loss via the distal skin. These procedures accelerated sleep itself and also increased delta activity and N2 sleep.

The principles involved in the aforementioned studies, although leading to specific treatments in future, for now emphasize how the environment and its management are crucial.

At a basic level, a room thermometer can be bought for only a very small outlay. It is actually essential to not simply rely on (the usual) hall thermostat, but experiment to improve understanding of what the relationship between the bedroom and the overall thermostatic temperature regulation means. This would only need to be adjusted for different seasons, so does not become an onerous task. Indeed, with the modern computerized thermostats, an hour or so of programming, is all that is need for the full year. Guessing bedroom temperature, if interested in a good night's sleep, is not helpful. If there is a sleep disorder or a brain injury lowered room temperature becomes critical, as I have demonstrated the lower the temperature the more likely SWS and healing will occur. This should come as no surprise to those that know anything about early trauma treatments and artificial brain cooling devices – now commonplace.

Similarly, a basic assessment of the quality of bedding, pillows, duvets and most especially mattress should be made. These are largely idiosyncratic but should, once again, be invested in. It is a marketing myth that one needs to replace a mattress every eight years; it depends entirely on the quality of the mattress, its usage, and most importantly: what one thinks each morning and the resultant opinion of the need to replace it. Whilst it is very likely a mattress costing tens of pounds will need replacing after a few years, one of the handmade ones costing many thousands, have guarantees lasting 10, 15, and even 25 years. A really expensive Hypnos, Vispring or Hastens mattress cost between £1,000 and £125,000. Careful attention should be made to the purchase at least of a summer and a winter duvet set, to keep the temperature optimized throughout the year.

Once again, having said all of this, as one of my hypersomnic patient's wives said to me: 'He could fall asleep on a clothes line and then not wake up 11 hours later, if the bomb went off!' A repetition of the dictum that diagnosis is crucial is apposite here. What her husband needed was the plethora of treatments discussed for hypersomnia, not necessarily a good mattress. Having said this it is equally critical to gain small improvements whenever they become available. In that sense, all of the above apply to OSA and hypersomnia, because they will potentially improve some of those elements critical for healing (e.g. temperature and deep sleep).

Noise, 'good' noise, no noise

Whilst it seems patently obvious that noise at night would wake a person, or at least disturb them in some way, our own personal experience cements this view. From external noise, the intermittent unpredictability of a partner's snoring, later perhaps a car alarm going off in the small hours, to finally an early morning shouting, clatter and bump of the dustbin or refuse collection, noise, appears unavoidable. All or even one of these noises can ruin a perfectly adequate sleep. However, scientists have made detailed efforts to delineate these much more clearly, especially at lower levels. The rationale is that subconscious effects may have distinct effects even though the sleeper is unaware of them during the course of the night.

For some time noise has been considered a problem for all aspects of sleep; from going to sleep, the quality of that sleep, to finally waking too soon. Most early studies recognized the overall problem but disagreed about the type, effects and most pertinently problematic methodologies. This is partly due to the individual differences in effects of the same noise beyond a particular level.

In this regard the WHO guidelines on noise and sleep published in 2018, for the European region, will provide a useful anchor to the following discussion. In examining three particular types of environmental noise (aircraft, train and traffic), and two others (wind turbine and hospital noise), the report concluded that each above 10 db background noise produced negative objectively measured physiology and subjectively measured self-reported sleep disturbances. Utilizing data from 2000 through 2015, the authors found the most robust evidence for a relationship between noise and cortical awakenings.

This examination has also led to specific examples of noise during sleep with particular health outcomes, including particular effects on the quality of sleep. For example, from an earlier period air traffic noise was demonstrated to increase sleep disturbances, awakenings, increased sleep medication and decreased SWS (Perron, et al., 2012). Indeed, these parameters have been supported robustly since this time. One of the better designed recent studies was the Discussion on the Health Effects of Aircraft Noise (DEBATS) one conducted in France. The results demonstrated effects from aircraft noise on objective and subjective measures of sleep. Sleep onset latency was increased, total wake time after sleep onset and sleep efficiency were both decreased. In addition, total time in bed was increased and reports of poorer sleep were consistently discovered. These judgements were arrived at for levels of noise and noise events. Equivalence of conclusions have been found across a range of different noise sources for level and event. For example: traffic (Pirrera, De Valck and Cludts, et al., 2010); trains (Licitra, et al., 2016); and wind turbines (Morsing, et al., 2018), amongst many other sources. Numerous polysomnographic and other measurement studies have confirmed these effects especially on two key sleep stages: REM and SWS or deep sleep (Basner, Müller and Elmenhorst, 2011). In addition, numerous studies have shown that after pain, noise is the most reported problem identified during and after hospital stays (e.g. Stewart, et al., 2017).

From a basic sleep management viewpoint, once again the idea would be to aim for a perfect monastic lack of noise in the middle of some 17th century rural idyll. Looking at this problem from a no noise perspective, one wonders what evidence is there that this is optimal for a good night's sleep. In their appropriately titled paper: 'Is silence golden? ... ' Kirste, et al. (2015) at Duke University found that indeed it was. Approximately two hours of silence promoted cell development in the hippocampus. Given this book's recurring themes of the hippocampi and their importance for neurogenesis and memory formation, this and related studies reprehensibly are not better known. Research springing from this in addition to the more well known studies highlighted which demonstrate the damaging effects of noise during sleep, has led to novel treatments including the resonant silence technique for those who suffer from dementia (see Fein, 2020). Fein's work, in crude summary, promotes silence following sound. This idea of 'intentional' silence, appears to have important therapeutic effects on anxiety, empathy, increased eye contact and even improvements in concentration, in a population of elderly people who have dementia. Moreover, numerous studies have shown that specifically modifying (through noise masking for example) noise, improvements have been repeatedly demonstrated through objective and subjective measures (Lee, et al., 2011; Demoule, et al., 2017). The latter study and another more recent one further cement the promotion of silence or at the very least the reduction of noise as highly attractive interventions for better sleep. In a study by Chaudhary, Kumari and Neetu (2020), earplugs and eye masks versus ocean sounds were utilized in a reasonably powered RCT to see which performed better. These authors found that earplugs and eye masks were significantly better than ocean sounds in improving sleep quality as measured by Richards Campbell Sleep Questionnaire (RCSQ) and simple patient qualitative reporting. The main flaw is obviously the problem of blocking two senses versus one. What this and similar work on white noise, other noise and specifically music has demonstrated is either no effect, poor effects (as with any noise) or only effects at the beginning of sleep prior to either disturbance or habituation occurs (for a good recent review see: Reidy, et al., 2021).

I would make a strong argument that anyone who has taken note of the summary of all the research and clinical evidence that has been provided, would find it odd indeed, nonsensical, to conclude any level of noise (even the supposedly effective white noise) was anything other than unhelpful to a good night's sleep. At its most basic, the ear has to process some level of this sound before shunting it further on to the auditory processing centres in the brain. In effect, you are not approaching the monastery, as at least one sense is still partially switched on; you are staying in the unhealthy 21st century. The only potential it has is for masking already noisy environments, such as shift workers who have no choice but to sleep in the much busier and noisier day period. Unless you fall into that category, please don't waste your money on any of these white noise (or similar noise based) products. Little or no research could be found examining these issues in the acquired brain injury population.

I would state a very simple but disabling symptom and note the commonplace of attentional filtering difficulties in this population. If you have problems in cognition, especially those involving information, sensory or attentional processing, the difficulties of noise are even more problematic. The only useful piece of work in this specific area was that by Elliott, et al., in 2018. They examined a group of TBI survivors. They concluded that broadly noise and light were problematic for a group who also suffered from PTSD due to chronic hyper-arousal. Of greater interest in this regard, is the fact that noise sensitivity was the second highest classification, when the group as a whole was stratified into what was disturbing them the most. This is the solitary piece of work found in this specific population and will have to suffice alongside what we know are more certainties previously described. Many years ago, I was privileged to hear the great Shakespearian actress Jane Lapotaire speak about the difficulty of simply being in a room with air conditioning, some years after her subarachnoid haemorrhage. She said that 'ordinary' noises that the assembled audience could accommodate to were hell for her and simply gave her a terrible headache. There is no logical reason to suppose that sleep, which is already problematic in this population, would somehow rectify these essential difficulties; permanently disturbed attentional filters don't suddenly correct themselves at night.

If soundproofing a bedroom is needed, do it. Unless fortunate to live in a supremely quiet rural area, some level of noise minimization is likely needed. The heavy curtaining I'm recommending in the next section will also deaden some outside noise. If double or even better triple glazing for a whole house, is too costly, at least invest in one of these for the bedroom window. If not, purchase some of the more modern ear plugs now available. These are now so light and gently malleable that they are essentially habituated to within a few weeks, and they cut any noise down considerably.

Let there be light (but only at the right time)

A theme throughout this book is that light, whether natural or unnatural, and of whatever wavelengths has a profound effect on both sleep and wake. As such, there is a great deal that could be written about this topic; indeed, a whole book should be devoted to it and its importance to humans. I will simply focus on those aspects that are of most importance for the purposes of basic sleep habits, management and then later, as it links with other foundations of good sleeping, post acquired brain injury: vitamin D3, the biome and the vagus nerve.

At its very simplest this is ensuring that the delivery of light and the obverse of the same coin, its blockage, are done at appropriate times. Thus, strong blue light of the sun, 10,000 lux light boxes, laptops, mobile phones and televisions are ideally delivered or bathed in during the course of primarily the morning, but certainly as a minimum, the day. As the day progresses into early evening the artificial producers of especially blue/white light in the shorter wavelength spectrum, should reduce in intensity and duration, as the natural light of the sun does. As the evening continues to progress into the night, at some point before bedtime no blue or white light should be present. During the course of the night complete and total darkness in the bedroom is the requirement.

If we break down the day in a little bit more detail consistencies will be found with previous discussions from earlier in the book. In essence, for the appropriately strong production of melatonin the shorter wavelength light should be available from waking and for at least the following two or three hours. Then 10–12 hours later, prior to bedtime, the somnolent systems will start to kick in. By this, if a light box is to be administered it is best to be administered, generally, (although remember my rule of the specific diagnosis), within the first two hours upon waking, for an absolute minimum of thirty minutes (ideally an hour). Following this, further consolidation of the shorter wavelengths of light can be enjoyed outside, if possible; as they have no real value behind a glass window. In this sense the ideal for television viewing, computer work on a bright screen or mobile phone usage should all occur during daylight hours. And as twilight emerges reductions in intensity of these natural lights should once again mimic the setting of the sun. During the course of the evening, as the darkness continues to gather pace these devices should gradually lessen in number to perhaps the last embers of the television thirty minutes to an hour before bedtime.

On going to bed the house should be lit with dim lights, optimally near to the infrared spectrum and the routine earlier described should also be done in such light (or much earlier in brighter lights). For example, some people wake up slightly during their teeth brushing, so do this earlier in the evening. Then upon entering the bedroom without a mobile phone or any other blue screen in situ, the final phase for going to bed can occur in a completely dark environment thanks to no external lights being able as full black out curtains have been fitted by a professional (this is important, for the price of a standard television, a professional can properly blackout a bedroom – why wouldn't you invest in one?) and the soundness of sleep and the regulation of melatonin can be assured throughout a restful night's sleep.

I have obviously caricatured the absolute ideal passage of a person through their light filled and less light filled day and complete dark night. Most people at some juncture through this would have giggled at the impossibility of such a prescription. One of the reasons for doing this is to think about an aspiration for optimal light effects during the day and the night. I realize that it may not be possible or achievable for some reading this, but that does not mean it should not be attempted. It also does not mean that if needed a small nightlight and some reading material is not available in the bedroom, as these are all part of the compromises we have to make in the 21st century to the electronic booming and buzzing world we live in. If I read the same paragraphs out to somebody in the 17th century, they would have thought it odd that I was simply describing their natural course of day and night bedtime arrangements.

Recent research has actually shown us the delicacy of the melatonin system. Before such knowledge, I would have probably pointed to three primary light and dark points in the 24-hour cycle. To begin this would be the essential first part described; getting an adequate burst of morning light in a sustained manner in order to kick start the full circadian programme. My second would have been to insist that people turn off any blue light product of any description a minimum of thirty minutes but ideally sixty before actually going to lie down on the bed for sleep. My third and perhaps as strongly as the first, I would have said your bedroom should be in absolute and total darkness. However, further research today on the subtleties of the interference effects of artificial lights has led me to worry that there are far more potential disruptors in this 24-hour cycle that we need to acknowledge.

What is perhaps surprising, is how little artificial light (10–20 lux) can disrupt the melatonin release at night, and suppress its proper function. It only takes two–five (2–5) minutes for this effect to occur. Early experiments demonstrated ordinary light in rooms (60 w and 100 w standard bulbs) caused delayed melatonin production in 99% of individuals and that delayed melatonin had a total duration of on average 90 minutes (Gooley, et al., 2011). In an interesting study which looked at nocturnal light exposure and its effects on melatonin and cortisol, found even lower levels of lux can affect these hormones. A continuous dim light of only just above one (1) lux had effects on these systems. However, the authors acknowledge that the brighter the light, the greater the effect, this is very clearly with participants asleep and eyes fully closed. Each randomly assigned light presentation, suppressed melatonin within 5 minutes and continued to the end of each light presentation, whatever the lux (Rahman, et al., 2019).

In a similar vein exposure to blue light products for sleep management programmes should take account of the specific problem of optical and ocular fatigue, as well as the more well known circadian disturbances. Ocular fatigue is more aggravated after blue light exposure, particularly LED devices (Lin, et al., 2017). Less obviously but potentially more damaging through its insidious nature, is the subtle disturbances of lower level blue/white light sources, at night, which can have profound effects on the human brain. Continuous exposure to low level (5–10 lux), mimicking that of the exposure in a normal street in a city significantly increases morning ocular fatigue when compared to total darkness. This problem of low level ambient light below 10 lux, has been repeatedly demonstrated, with participant's eyes closed, to cause a number of sleep quality issues. It decreases total sleep time, sleep efficiency, and increases in REM sleep along with concomitant reductions in NREM sleep, in addition to the previously noted ocular fatigue (see for example: Park, 2018 or Xiao, et al., 2020).

The point here is that as with the discussion on noise, even partial, ordinary level, non-direct or even very low level light has negative effects on the sleeper. It is almost as though anything other than total darkness has some form of negative effect and the further away from this the artificial light becomes (in terms of greater lux and duration), the more deleterious the overall effects. The most obvious are those that are quickly seen, felt or able to be measured through experimental research, perhaps at the end of a night or series of night's sleep. However, increasingly research is accumulating evidence surrounding the chronic effects of light disturbance. In a population based cohort study undertaken by Min and Min (2018), results demonstrated that night time levels of outdoor artificial light were associated with the daily dosage of hypnotic medications used by older adults. Even more pertinently the higher the level of such light greater the overall length of usage and the higher the dosage. Given my prior account of indicators of chronic inflammation in the older adult population, the increased rates of insomnia and the clear links between these and dementia, the weaker suggestion would be that such lighting exacerbates dementia, the stronger suggestion would be that it could be one of a number of causative factors. Rajput, et al. (2021), in their review of artificial light at night, in addition to the more well known and obvious sleep disruption, circadian disorders and obesity, also proposed this kind of light at night has clear links to certain cancers. Furthermore, they proposed at least one potential mechanism for resolving these, that of the 'wonder molecule' melatonin. In my view this is too simplistic and does not take account of all the other ideas I have discussed in the book, most pertinently the various increases in chronic inflammatory processes, which although can be ameliorated somewhat through melatonin, cannot wholly be cured by it.

In terms of artificial light and those who have suffered an acquired brain injury, highly specific effects have been found in animal, and human studies. In effect, these studies demonstrate highly detrimental effects on exactly those neuronal areas most commonly found to produce functional results. For example, in an experimental study, exposing finches to five (5) lux of artificial light for seven (7) weeks compared to a total darkness group, distinct differences were found in neuronal recruitment or degeneration; when there should have been an approximate 'dull parity' being the planned apoptosis described earlier. In effect, artificial light exposure increased this disparity and exacerbated responses from previously undamaged regions. This leads on to the body of human research, some of which has been cited, which demonstrates the inflammatory results of acquired brain injury and sleep disturbances through artificial light exposure (Kilgore, et al., 2020).

The conclusions and recommendations are easy to suggest and require minimal outlay in practice. Remove all blue light products from temptation: TV's, laptops, computers, smartphones, or any type of screen, or even a hint of a blue screen. Next would be professionally fitted blackout curtains in the bedroom, or good quality blackout curtains you've fitted yourself, or at the very least heavy dark curtains, and no lights (low level, ambient or otherwise) in the bedroom itself. Good quality cushioned night eye masks are a definite one down option, due to the sensory provoking problem, but they are far better than light of any kind. On other levels complain to your local council about LED lighting, it is harmful and nowhere near as 'green' as they are making out. If they insist on its presence, request a yellow or amber filter to reduce the blue light glare. Light pollution is an enemy of all animals and our planet; moving to cheaper LED street lighting is not an option. Indeed, this form of lighting causes significant harm to our sleep and is harming other species.

Naps and weekend extras

The most common question around these instructions, after I deliver them, is either can we nap and for how long, and if the routine should alter dramatically over the course of a weekend. The answer is, it depends on which data one pays most attention to for this purpose. Apart from this there are general rules that are very useful to follow.

First, napping. This depends on where you are in your recovery as someone who is recovering from an acquired brain injury. It is generally advisable to take three (3) to six (6) month tranches as useful from the point of injury, in addition to the severity of the injury and location. In this regard, napping is actually recommended for at least the first year, in the afternoon. It should be most evident and time consuming in the first three to six months, indeed if someone recovering from a brain injury is not napping during this period our experience is usually that the outcome is poorer. I can find no studies to support this view, and intend to rectify this shortly for those with an acquired brain injury, but it is a very strongly held view across all the members of our service and other similar services. It is also a useful barometer of recovery; in that significant napping in these first six months, if all is going well, is largely replaced by less frequent and less prolonged napping as time goes on. It is usually after the first six months that proper cognitive testing, and diagnoses around such things as hypersomnia become easier. The next three–six months recommendations are thus titrated more clearly and allow for further indicators of progress (or sadly otherwise) to be demonstrated. For cortical stroke recovery the majority are at a point after approximately 12–18 months, where most if not all napping is unnecessary. For most sub-arachnoid haemorrhages, and hypoxic injuries, the yardstick is more likely 18–24 months. However, TBI will take longer and probably the yardstick here is more likely 36–48 months.

Other than these general ideas, there is considerable evidence for the support of 'good and bad napping.' First, it is well known that late afternoon napping; generally held to be after 4 pm, is very likely to interfere in some way with the night's sleep ahead. In this regard, most research supports the view that afternoon napping, as long as it is between the hours of 1 pm and 3 pm can be beneficial, with certain caveats, which will be outlined below. In addition, numerous studies have demonstrated the health, cognitive and emotional benefits of napping. It has been shown to be able to boost memory consolidation and future learning (Antonenko, et al., 2013, Baran, Mantua and Spencer, 2016), executive and frontal functioning (Hayashi, Motoyoshi and Hori, 2005), emotional regulation, and most pertinently for one of the primary themes in this book, anti-inflammatory processes (Faraut, et al., 2015).

Yet, there is also a wealth of research that suggests, especially in later middle age and older adults, frequent and or prolonged napping is linked to negative health, including but not limited to: diabetes, depression, hypertension and increased mortality. Moreover, it is a very good predictor of cognitive decline; Cross, et al. (2015). This latter napping has been described as: 'essential napping' and has been suggested to be dependent upon the general state of chronic ill health, which produces so much inflammatory problems that the person is 'commanded' to attempt to mop these up through prolonged sleeping and additional daytime napping. This need for daytime napping in the older adult population is based on a similar need (or a Hebbian drive?) in the brain injured population, it is notable that both groups have increased nocturnal and daytime napping that are both NREM rich. This idea is also consistent with research that suggests napping (in addition to the more obvious nocturnal sleep) resets, or at least attempts to reset, inflammation levels back to normal (Faraut, et al., 2011). As with earlier, I come back to the idea of some inflammatory responses are essential, especially in the acute phase of recovery from a brain injury but chronic inflammation is too harmful to cope with.

There is a wealth of research that supports the old Spanish and Greek habit of an afternoon siesta, after eating lunch and the lower incidence of cardiovascular problems for those that do so (a recent example being research published in *Heart*, 2019, by Häusler, et al.). In this sense the nap at the correct time, and frequency, is potentially protective of inflammation and later disease. Napping in older adults after any meal is a way for the body to attempt to mitigate existing inflammatory processes in those that are likely to have existing disease. Moreover, nap duration is key depending on requirement. Thus, it seems a 10-minute nap is optimal for immediate alertness and increased cognitive focus for the short term. However, it is less good at resetting inflammatory processes. A better option for this latter need seems to be a 30-minute nap, as this allows at least some minutes in SWS, which of course switches on the glymphatic system of cleansing. However, it takes longer to become alert after such a nap, so needs to be carefully planned, in order to maximize benefit. Although, as one may predict, it has longer lasting benefits than the 10-minute nap. After the initial sleep inertia from waking, longer alertness levels kick in and last for a greater duration (Hayashi, Motoyoshi and Hori, 2005). Clues as to why these functions are more pronounced in the acquired brain injured population can be found in an interesting paper by Stern and Naidoo (2015). They found that wake promoting cells (e.g. histaminergic, orexinergic, noradrenergic) are depleted in those with increasing age and this effect was more pronounced in those who in addition, had neurodegeneration. The hypothesis would be that those with reduced brain integrity and overall brain functioning had a greater propensity for sleep and thus daytime napping. However, that reduced brain integrity increased the chronic inflammatory processes previously highlighted together with their potential mitigation mean increased need for conscious napping is unarguable.

It may be that the link with increased morbidity and mortality surrounds the acute/chronic dimension as well as the quantitative problem of the body trying to cope with this, repeatedly seeking napping or extra hours of nocturnal sleep, but failing to do so. These dimensions would not only explain the differences found in the research with age, but also with the exact length of napping, and either its positive or negative outcomes. In other words, both of these research streams are accurate, but the explanations needed to understand them are different. Thus, for too long argument has raged about napping being either good or bad, when in actual fact it can be either at different time points in a life, at different time points after a brain injury, for different times of the totality of a nap, and different times during the day. I feel there may well be a theme here.

Weekend catch up?

Finally, in this section, a few words on weekend catch-up sleep, or 'lie ins.' Unfortunately, the answer here is both yes and no, but it depends on how well you sleep in the week. In taking the no answer first, if one was a strict adherent to the behavioural approaches, that I'm a strong supporter of, then one should always try, apart from the very few occasions in life (New Year celebrations, weddings and so forth), that may warrant a break from rigid sleep routines and structures.

If each and every weekend a two, or even three, night alteration to this schedule arises, then the very essence of the neurobehavioural shaping principles will be undermined. Then at the start to the week, each night will take time to re-accommodate to the 'normal' sleep routine. So the conclusion here would be no, do not do it. Historically, it was also felt that you cannot 'catch-up' on missed sleep during the weekend. It was gone, never to return. So, again the idea of catch-up has little or no support; that is except for more recent work. Much publicity was given to a Korean study on weekend catch-up sleep and a later Swedish study. In effect, the Korean study; Oh, et al. (2016) was a large survey project that sent a sleep questionnaire and a quality of life questionnaire to 4,871 participants. The results, at least understood through quality of life (QoL), was it was higher in those that undertook weekend catch-up sleep, this effect was markedly so for those that averaged seven hours or less each weekday night. Thus, if you are an individual who could be described as outside the recommended window for most adults (seven–nine hours), then there is at least a subjectively significant positive effect in having a weekend catch-up sleep. Even more important for the overall debate was the later study reported by Åkerstedt, et al. (2019) in the *Journal of Sleep Research*. They examined a very large cohort of 43,880 Swedish people, and followed them over 13 years, through record-linkages. Standard Cox proportional hazard regression models were then used to examine the data. The conclusion is potentially predictable from what I've written thus far, with a slight twist, they stated: 'In conclusion, short, but not long, weekend sleep was associated with an increased mortality in subjects <65 years. In the same age group, short sleep (or long sleep) on both weekdays and weekend showed increased mortality.'

In other words, if you're too far outside the general 'good' parameter for effective sleep (seven–nine hours), either too few hours or too many hours, then there are clear negative health effects. In addition, shorter sleep at the weekend was a further negative effect. Indeed, they found an approximately 65% higher risk of early death in those that slept for less than five (5) hours each night on average, when compared to those who slept between six (6) and seven (7) hours on average. Moreover, some increased catch-up sleep had some positive health effects, even in the lower average group (they just were not as profound), but there were none found in the over 65 age ranges. In other words, it is better to have some weekend catch up sleep, especially if you have little average sleep on a weekday, it may well save your life. This only applies if below 65 years of age.

Exercise

Exercise has been left to the end, as it is rather a special case, for it is the only known (so far), intervention that not only enhances sleep but improves the glymphatic and meningeal lymphatic systems. In other words, it is the only thing we have to improve the key problems of waste disposal. In some respects, it could have been a fifth foundation. That it was included in the largely behavioural first foundation, was deliberate. It is essential that exercise becomes a part of the wake routine, it is only through this, extra waste disposal can occur, allowing a slight easing of pressure on the damaged systems operating at night. It cannot be an option, but it must also become, as with teeth cleaning and a clear bed and wake time, a cast iron, scheduled, unalterable part of a person's routine.

First, exercise should be limited to no later than four (4) hours before bedtime. It is also true to say that regular exercise assists in sleep and the overall reduction in morbidity. It seems to have a synergistic effect on other health related behaviours; so it should be seen as integral to a good sleep management programme as well as obviously benefiting the person overall. One of the ways morning or afternoon exercise helps is that it appears to either reset or cement aspects of the circadian rhythm. In particular, this is done by raising body temperature earlier in the day which appears to pro-voke better body cooling needed later on, this effect is further enhanced through outdoor exercise. It is a myth that later exercise causes sleep pro-blems, whilst this is true for most; a significant number are not affected by early evening exercise. It is very important to experiment and know the body; however, the four hour rule applies to most. These positive exercise effects have also been demonstrated for those with a brain injury (see for example: Bogdanov, et al., 2017; Ding, et al., 2020). As this harmonizes with the instruction for avoiding large meals this should be easy.

Once again, an apparently simple intervention is actually a very powerful one. Numerous studies in mice have demonstrated the importance of vo-luntary exercise and improvements in glymphatic influx and (critically given the earlier discussion) efflux (He, et al., 2017, von Holstein-Rathlou, et al., 2018, Liu, et al., 2020). Much of the more recent research efforts in this area, having proved the efficacy of exercise on the glymphatic system have now moved on to more specific understanding. One such stream has focused on if there is a critical time window, after which exercise no longer benefits. The research by Liu and colleagues (2020) found that exercise was dependent on a well functioning AQP-4 water transport component of the overall system to produce preventive results; and this was found only in younger mice. Once they reached seven (7) months of age exercise lost its power. Research parallels that undertaken in animals then subsequently in humans on car-diovascular disease. This ultimately found unless exercise is undertaken prior to the early 65, the heart has lost too much of its elasticity to benefit. If exercise occurred prior to this, it was shown to significantly improve cardiac stiffness, and that this effect was lasting. However, the heart became too stiff to be helped as significantly after this age (Howden, et al., 2018).

At this point we have research in mice that demonstrates equivalent possibilities in the glymphatic system and neurodegenerative disease. Another stream of research has focused on the immediate benefits of ex-ercise on the glymphatic system. von Holstein-Rathlou, et al. (2018) found that mice who exercised during the day had twice the overall glymphatic activity, than mice that did no exercise.

I am not waiting for the confirmatory evidence in humans, prior to re-commending it. Exercise has been repeatedly shown to be of benefit in re-ducing neuro-inflammation. In the course of time, as research accumulates it may become a fifth foundation, or a new and better approach to enhancing the waste disposal systems will be found. For now, it can be made a strong recommendation for the purposes of this book, as part of a good daytime routine. It also perfectly illustrates a repeated theme that the day is as im-portant as the night, good sleep happens because the right things happen in the day. With as much certainty good days happen because the right things happen at night. The frequency, intensity and duration of such exercise is dependent on the patient. There is no one size fits all in this process.

Chapter 12

Second foundation

Writing this, as I am, during the time of the COVID-19 pandemic, it has been widely debated as to the necessity or otherwise of increasing Vitamin D and thinking more seriously about our gut biome than we had ever had recourse to do before. What may surprise readers is that in order to ameliorate any sleep disorder, but particularly the three primary disorders that are most common after a brain injury (insomnia, hypersomnia, and sleep apnoea), Vitamin D and our gut biome are critical for optimal sleep functioning.

To begin, some thoughts around Vitamin D. It is easy to confuse the fact that there are a number of Vitamin D forms that have been discussed in the literature. Even more interesting is how the one most specifically of importance for the purposes of this part of the book, Vitamin D3 is actually a hormone, which was only discovered in terms of its chemical structure as recently as 1935. Usually, when Vitamin D is left without a sub lower number; it is taken as either the D2 or D3 forms, thus from now on D will be used, unless specifying supplements. In varying degrees, the collection of D vitamins from one to five are chemically most like steroids in terms of composition. Indeed, it is classified as a secosteroidal hormone. Testosterone, cortisol, oestrogen, and melatonin are biochemically all very close steroidal relatives.

Crudely, Vitamin D is formed largely through the skin's exposure to sunlight, the chemical substrate of which is metabolised initially by the liver. It is then finally turned into the hormone Calcitriol (also known as: 1,25-dihydroxycholecalciferol) by the kidneys, which is the active form and as such produces multiple effects over multiple locations and systems within the body. Much of the problem in understanding this complex hormone is the reliance on partial data produced shortly after it was discovered in 1935, which concerns its potential ability to modify calcium and, as a consequence, childhood Rickets, and more generally ideas around bone health. However, in this regard and perhaps ironically, there is no good evidence that Vitamin D supplements actually prevent osteoporosis. In one of the most recent reviews, Reid and Bolland (2020), concluded that supplementation

DOI: 10.4324/9780429199066-12

cannot be recommended, for that reason. For our purposes, the most important thing to understand is that more recent, high quality research, meta-analyses and systematic reviews have demonstrated, over the last two decades in particular, robust associations with Vitamin D deficiency and a variety of sleep disorders.

In this regard, one of the better, more recent systematic reviews was undertaken by Gao, et al. (2018) which would provide the reader with a useful summary of the current state of play. This summarised a number of studies involving over 9,000 participants. They found that all participants with deficient Vitamin D had a substantially higher chance of disordered sleep. Further analysis into subgroup data showed that Vitamin D deficiency was specifically associated with poor sleep quality, shorter sleep duration, and increased levels of daytime sleepiness. This analysis also indicated that this was largely found for those who had <20 ng/mL of Vitamin D. This has been confirmed by a number of other recent reviews and studies (Kim, et al., 2020).

Moreover, it has now been demonstrated across the age span. Al-Shawwa, Ehsan and Ingram (2020), in a retrospective cohort study, concluded that Vitamin D deficiency in children was associated with objectively measured, decreased duration and poorer sleep efficiency. Their study, which involved laboratory-based polysomnography and measurements of serum Vitamin D, had the more generous cut off of <30 ng/mL. Interestingly, they also discovered that Vitamin D deficient children were more likely to have delayed bedtimes and in a related part of their analysis questioned the potential for a relationship with this hormone's effect on melatonin; potentially reducing its secretion. This of course is unsurprising to those who have paid attention during the course of this book over my concern about the primacy of melatonin and its relationship to a wide range of different factors. However, even very high supplementation of Vitamin D, does not appear to affect melatonin, in any appreciable degree. It may do theoretically however, and as such, one recent review has suggested better anti-inflammatory effects may be gained by providing both supplements together (Marón, et al., 2020).

However, if we briefly come back to the fundamentals of the importance of both sleep and wake, it is largely expected given that Vitamin D is mainly manufactured during the course of sunlit hours in the daytime, it is integral to melatonin and has a complex relationship with it (Golan, et al., 2013). Vitamin D also affects sleep indirectly. One of the interesting ways that it can affect sleep is its relationship with ferritin (iron). It is important to note that for a large subset of those who suffer from restless legs syndrome (RLS), iron supplementation provides relief and, in some cases, complete resolution of RLS. However, it may be more sensible to offer a Vitamin D supplement of the right kind and dosage on a daily basis in order to rectify the potential iron deficiency problem. This has now begun to occur in the literature and a body of research is beginning to emerge on Vitamin D's benefits for RLS, however the mechanisms are thought at this point to be multifactorial. Wali, et al. (2018). Their population-based case control study hypothesised that in addition to Vitamin D's effects on iron within the body, a more primary concern was its known effects on the dopaminergic system and whether deficiencies could be robustly demonstrated in those with RLS. This was indeed found to be the case. In other studies by Wali and his colleagues, similar positive effects on RLS were found (2015). The mechanism is thought to be Vitamin D's ability to down regulate pro-inflammatory cytokines and then how this affects the bioavailability of iron (Masoud, et al., 2018).

In another related study, specifically looking at qualitative aspects of sleep, Majid, et al. (2018) in a randomised control trial looked at the potential for Vitamin D supplementation on those who had diagnosed sleep disorders based on the Petersburg Sleep Quality Questionnaire (PSQI). The PSQI score reduced significantly in vitamin recipients as compared placebo. This difference was significant even after modifying variables were analysed. Specifically, in the 20–50 age group studied, vitamin supplementation improved sleep latency, sleep quality and sleep duration in those who had a previously diagnosed sleep disorder. Someone in the future would look back on our 20th-century obsession with Vitamin D in regard to simply bone health and chuckle at the ridiculousness of our *not* being obsessed with Vitamin D and its relationship to sleep.

However, it does feel like the debate regarding Vitamin D and immune function as regards COVID-19 is an unhappy parallel, as at this point it has now been demonstrated that Vitamin D, if given *prior* to somebody becoming very ill with Covid, is protective and indeed reduces overall morbidity and mortality by a considerable degree. Not only this a very recent updated meta-analysis in the *Lancet*, demonstrates a barely adequate supplementation (very small doses of 400 *iu*–1,000 *iu*) offers protection from developing acute respiratory infections of any kind (Hernandez, et al., 2020; Joliffe, et al., 2021). What this implies is that, once again, at the proper dose it could have even greater effects on a number of processes. At least half of the difficulty with this problem has been that older trials undertaken with Vitamin D supplementation would come under the heading of 'poor' simply because they attempt to either provide the wrong type of Vitamin D supplement that is difficult for the body to absorb and/or they administer this in gigantic doses over the very short term and then conclude it has not worked. All those studies that were sensibly designed around the idea of mimicking Vitamin D delivery so that it is more like being in the sunshine on a daily basis have demonstrated the biggest effects in outcomes. Supplementation of Vitamin D does not work instantly and takes a number of days, weeks, or even months depending on the deficiency of the individual, and the kind of Vitamin D delivered, for it to come up to required levels. It should be noted very strongly at this point, that any study wishing to demonstrate a similar effect on sleep should have the right form of Vitamin D3 (e.g. cholecalciferol) or its product and a steady amount needs to be delivered on a daily basis until the correct levels are assured through blood testing.

Coming back to Vitamin D and sleep specifically, what is currently known is the drive away from the consensus of understanding Vitamin D deficiency and numerous metrics of sleep problems. That would appear unarguable. From such up-to-date consensus, cutting edge science has now begun to look at Vitamin D supplementation, bright light, or increased sun exposure in order to rectify these sleep problems.

As a hormone with wide-ranging effects, which have been demonstrated through those who have adequate amounts of vitamin D and those that are deficient, it seems that beyond the very robust and repeated associations with Vitamin D and its effects on sleep, we are now moving towards a deeper understanding of the components of this. One such is how Vitamin D is involved in the most prominent pathways in the production of melatonin. In addition, it has an indirect and a direct ability to reduce neuro-inflammation. As such, it would also appear to ameliorate non-specific pain disorders which correlate with sleep quality, to even more recent claims that it may reduce the severity of such a supposedly obviously mechanical disorder as sleep apnoea (Siachpazidou, et al., 2020). Moreover, Vitamin D receptors have now been discovered in the same regions in the brain that regulate sleep, it also appears to be involved in a more direct way (outside its importance for melatonin) with regulating the sleep-wake cycle (Muscogiuri, et al., 2019).

In an interesting study from Brazil by Piovezan, et al. (2017), utilising a cross-sectional approach 657 patients were assessed using a number of measures. The participants were examined using clinical interviews, standardised questionnaires, blood sampling and polysomnography measurements. Obstructive sleep apnoea (OSA) was then classified from this as mild, moderate or severe and another key aspect of the research was to identify, in addition to any relationship with Vitamin D and OSA, whether there was a relationship between Vitamin D and short sleep duration for less than six hours. The results clearly found that OSA and shorter sleep duration showed substantial correlations with Vitamin D deficiency. These findings were most prominent for those over 50 and who had greater severity of OSA. The authors concluded '...Age-related changes in Vitamin D metabolism and the frequency of sleep disorders may be involved in these associations.'

A different type of intervention study by McCarty (2010) discussed a clinical case of hypersomnia in a young woman, who underwent a full polysomnographic examination and was also clinically evaluated and standardised measures were administered. The conclusion from all of these was that no features were characteristics of sleep disordered breathing (e.g. OSA), no features of depression or narcolepsy were present. However, significant daytime fatigue and related hypersomnia was diagnosed, in addition to pain in her back as well as her thighs. In addition, she suffered chronic daily headaches, and her Vitamin D levels were exceptionally poor (measured at 5.9 ng/mL). The only single treatment initiated was high dose Vitamin D supplementation: her hypersomnia improved within two weeks. At later follow up she had a clear polysomnography test result, and all her various chronic pain problems had completely resolved. Her Vitamin D levels were remeasured at this point and found to be at 39 ng/mL. The author concluded that the Vitamin D supplementation reduced non-inflammatory myopathy, modified the release of tumour necrosis factor-alpha (TNF-α), in addition to the more well-known and obvious modifications in melatonin and inflammatory modulation associated with Vitamin D. We are thus moving away from the idea of the importance of Vitamin D and its relationship with sleep towards interventional studies based around supplementation. A number of studies have shown that Vitamin D supplementation improves quality and improves sleep duration in patients with sleep disorders. Other studies have found less consistent results (Gunduz, et al., 2016). My own view is rather like the studies discussed earlier in regard to COVID-19; part of the reason for the inconsistency in findings is predominantly due to either the type of supplementation delivered or the timing and dose of delivery. A general trend could be found in the literature for improvements if supplementation was delivered in the correct form on a daily basis. Less helpful results were found when the Vitamin D supplement was a harder to absorb kind and/or it was delivered in an infrequent large dose. A helpful meta-analysis highlighting some of these issues has been undertaken by Yan, et al. (2020).

In addition to the well-known effects on melatonin and sleep regulation, Vitamin D has also been proposed to have significant effects on inflammatory processes in the body. A good illustration of this is the work undertaken by Mansournia, et al. (2018), who undertook a systematic review and meta-analysis of randomised control trials. It examined the relationship between Vitamin D supplementation and various inflammatory and oxidative stress markers. Their conclusion was that Vitamin D supplementation had profound effects on reducing high sensitivity C-reactive protein and malondialdehyde (MDA). Moreover, this same supplementation was found to increase markers helpful to good cerebral and vascular functioning, including increased nitric oxide release, total glutathione, and total serum antioxidant capacity. This begs the question of independence from the effects of melatonin; which I think is not possible to know at this point. Separating the inflammatory effects of either, or both would be problematic. Given the previous emphases in this book, it may not surprise the reader that all of these strands point to the great potential of Vitamin D as a treatment for those people who have suffered a brain injury and who have concomitant sleep disorders.

Parallel findings for Vitamin D deficiency in the ABI population point to a similar pattern of difficulties. For example, Toman, et al. (2017), in their single centre prospective observational study of a traumatically brain injured population, looked at the prevalence of Vitamin D deficiency and the relationships between this deficiency and either severity of brain injury or quality of life. Multivariate regression analyses were performed, controlling for age, season, ethnicity, time since injury, gender, and TBI severity at point of injury. They found a relationship with Vitamin D deficiency and severity; those with a severe TBI had significantly lower Vitamin D levels than those with a mild TBI. Self-reported quality of life was better in patients with optimum Vitamin D levels than those with a deficiency. They concluded that clinicians should actively screen for and subsequently treat Vitamin D deficiency in the head injured population. This team are now undertaking further research to identify at which point after an injury Vitamin D deficiency develops, and potentially whether Vitamin D replacement benefits recovery and quality of life. Similar findings have been demonstrated in other teams, for example Dubiel, et al. (2019) found that in a much larger cohort than the previous study, the majority of patients undergoing acute inpatient rehabilitation were found to either have insufficient or frankly deficient Vitamin D levels. Only 39% of the population that were studied were found to have adequate levels of Vitamin D.

Vitamin D deficiency should be understood as a problem amongst all those who suffered any kind of a brain injury, rather than just the traumatically brain injured population (Jamall, et al., 2016). An extremely good recent example can be found in the systematic review undertaken by Yarlagadda, Ma and Doré (2020). Their view was that the research to date demonstrated that Vitamin D deficiency is a significant risk factor for ischaemic stroke and should be added to the more commonly understood list of risks for developing this problem, such as age and genes. Moreover, and as profoundly they asserted that the same literature found that the severity of the initial stroke and all acute and chronic outcomes were made worse through Vitamin D deficiency. They then propose several potential mechanisms and pathways as to why Vitamin D might potentially mitigate stroke onset and the severity of outcome. Interestingly, as I have previously examined, one of their core proposals is around the idea of its mechanism in inhibiting reactive oxygen species and other neuroprotective factors, which may mitigate some of the other inflammatory processes outlined. They then conclude that there may be some small evidence that Vitamin D supplementation could lower the risk of stroke and improve recovery, but the latter results are quite mixed. I am reminded of the discussion in regard to optimal dosing, timing, extent, range, and type of Vitamin D earlier.

Prior to unpacking these elements further, it is worth noting that the literature review undertaken by Anjum, et al. (2018) proposed further mechanisms for the effects of Vitamin D in the required amounts and when deficiency of it occurs. Specifically, for aspects of brain functioning, they suggest that Vitamin D has a vital function in neuronal and glial tissues and the effects of its deficiency can partly explain the difficulties through a lack of the buffering of the required calcium in the brain. This deficiency leads to increases of cerebral soluble and insoluble amyloid-beta (A-β) peptides and a concomitant decrease in the brain's anti-inflammatory and antioxidant properties. Once again demonstrating the link with these patterns that have been repeatedly implicated in poor brain health associated with a wide range of conditions, most obviously dementia. My own conclusion, looking at the available literature, is that it is relatively easy to support the case for increased problems in ABI partly as a consequence of Vitamin D deficiency more readily up to the point of infarct or other insult. However, there is somewhat less robust support for ongoing problems associated with Vitamin D deficiency that could potentially be propelled by the injuries themselves. This also makes the severity statement earlier moot. Thus it may be that the greater the severity of TBI the larger the reduction in Vitamin D, utilised by the body to heal. In similar fashion to the mismatch of melatonin earlier and later post injury. One of the reasons for this could be the way this hormone, if available, is the main producer of both β-defensin and cathelicidin. These are very important mediators that can be switched on both by Vitamin D and certain probiotics (Plaza-Díaz, et al., 2017). β-defensin and cathelicidin are critical for switching on the main anti-inflammatory pathway (through IL-4, IL-5), in addition, through this process and the subsequent release of IL-10, they turn down the major pro-inflammatory cascade leading to increases in IL-6, to take one prime example. Thus even if the TBI leads to depletion of the hormone, its importance for exactly the kinds of problems previously discussed is crucial. It must be a case of how best can we ensure proper amounts of Vitamin D stocks are replenished in this population; not whether its important.

Research into the supplementation of this hormone and other forms of treatment by it in the ABI population has only really developed in the last ten years or so. The stock of knowledge is therefore tentative and not conclusive. One such example of the modern approach to utilising Vitamin D supplementation can be found in the work of Yang, et al. (2021), who studied the different dose effects of Vitamin D via intraperitoneal injection one week before induced TBI and three weeks after in rats. Standardised neurological assessments were taken. They concluded that supplementation of Vitamin D ameliorated neurological deficits and cognitive impairments induced by the TBI. Furthermore, and most pertinently for our purpose, it reduced overall brain oedema and decreased inflammatory responses. Critically, the latter showed significant reductions in cytokine inflammation and decreased overall blood-brain barrier (BBB) disruptions. Although somewhat more wide-ranging, a case series has been presented by Matthews, et al. (2013), which demonstrated reversal of coma and improvements in clinical outcomes in patients with very severe traumatic brain injuries, utilising Vitamin D, progesterone, Omega-3 fatty acid and glutamine. The conclusions were so positive, a larger clinical trial is now being investigated. The momentum for supplementation studies has now gained further impetus. This is witnessed by a different new study protocol, which is investigating a randomised placebo-controlled trial of high versus low dose Vitamin D supplementation. For serum levels of inflammatory factors and mortality rates in severe TBI patients by Arabi, et al. (2020). The examination will look at inflammatory markers, such as IL-6, monocyte chemoattractant protein-1 and C-reactive protein in non-critically ill patients, and will hopefully be reporting shortly. Other markers will be looked at including those factors that may be associated with Vitamin D deficiency or sufficiency including comparisons of infection rates, time of discharge and any incidences of delirium.

Time to talk about more hormone treatment

In my view, Vitamin D, both as a supplement and as a minimum recommendation for daily production via sun exposure, is therefore the second foundation of not only treatment for sleep, but also a support to the anti-inflammatory needs of the sleep deprived and the acquired brain injured populations combined. This straightforward advice is complicated by a number of factors, most prominently, it begins by understanding where you live in relation to the sun i.e. latitude. Thus, on a sunny day at the end of June in the northern hemisphere, whether you are in New York, London, or Paris, you may expect to make the maximum that you need, of 20,000 international units (iu) of Vitamin D. In order for this to occur, you would need to be in the sun with at least 25–40% of your body exposed, as a minimum trunk, arms, and legs. At this point in the year, you only need to be out in the sun for 30–60 minutes to achieve this level of Vitamin D. Obviously, anything either side of this day in high summer and anything less than a perfectly sunny day produces decreasing amounts of this hormone. It is difficult to pinpoint hard and fast rules, as so much differences are apparent with latitude, skin colour, weight, time of day, the season, and cloud cover, pollution levels, amongst other things. Various sunshine calendars are available online to check more precisely the amount of potential Vitamin D you may make over a given period of time at a particular point in the year, in your particular location.

Having said all of this, various other factors need to be borne in mind. First, a person's weight is significant for the amount of Vitamin D freely available in the body, thus the larger the person, the less available to begin with as Vitamin D is first absorbed by a person's fat cells before becoming available to the whole of the rest of the body. Secondly, skin colour is important for how much exposure produces what amount of Vitamin D. An example is that someone with very light skin may make enough Vitamin D in the space of 10–15 minutes, whereas the equivalent may not be made in a darker skinned person within 45 minutes to an hour. All these factors need to be considered alongside your particular level of latitude. Put simply, someone in London may have to spend 90 minutes for the equivalent return of Vitamin D as someone in Honolulu for 30 minutes. The other implication of the research that we have available is that supplementation is necessary for anyone of any skin colour and any latitude through the darker winter months, and the darker and longer those winter months are, the higher the supplementation needed. In other words not enough Vitamin D will be produced at certain times of year, moreover it cannot be adequately provided by our diets. That is unless you are an Inuit living near the Arctic circle eating daily seal and whale fat!

The critical thing is to measure it adequately. General practitioners should be able to measure it adequately, however if there is a difficulty, as this research may be unknown to them, at home test kits are widely available and as long as they are attached to a reputable laboratory for assessing the results, can be turned around in a few days, with a sample of blood (with a small lancet provided) sent in the post. According to most standard authorities, the minimum requirement is a blood level of 30 ng/mL (which is the equivalent of 75 nmo/L). Anything below this level is considered by those, such as the FDA in America and the NICE guidelines in the UK, and other similar authorities across the world, to be deficient. However, for optimal functioning outside of bone health (*sic*), both functional medicine and certain colleagues in neurology and endocrinology (Dr Gominak and Professor Holick) suggest the optimum levels are actually 60–80 ng/mL (150–200 nmo/L). This means that the supplementation necessary in the darker d ays of winter is more likely to be 5,000 *iu* per day between 1st October and 1st March, approximately. At this point, it is worth noting that there is a small percentage of the population that would need much lower doses due to a condition which leads to potential hypercalcaemia. This needs to be known before setting off on the supplementation journey.

All of the above leads one to conclude that, whilst increased amounts of Vitamin D are essential, for brain injured patients, before embarking on an increase of this hormone, the blood test is fundamental. Indeed, during the course of the first two years of titrating this, one should aim for four tests each year, one during each of the seasons available. In this way, safe increases to the levels required could be achieved. At this point, you will have a deep understanding of how your body works with the sun at different times of the year, at different latitudes, with different levels of supplementation, in order to keep the overall level optimal. Most frequently, what is needed is a period of boosting and then a slightly different level of sun exposure and supplementation throughout the year in order to maintain the correct level. Hopefully this will result in a low to no dose supplementation between May and September (in Northern latitudes), due to an adequate sun exposure routine. This is unquestionably the best way to manufacture this, and the exact rate of supplementation would vary in the late Autumn, darkest winter, and early Spring time, again depending on skin tone and exact place on the earth. Between October and April supplementation becomes essential for those in the Northern hemisphere.

Chapter 13

Third foundation

From a hormone to the gut biome

As I demonstrated, Vitamin D is an important hormone which affects a wide range of functions either directly or indirectly. Recent research has found that Vitamin D receptors are highly available in the gastrointestinal tract, affecting gene expression. The particular importance and relationship between Vitamin D and our gut biome has long been debated, but recently, researchers at the University of California have found that there is a significant association between greater diversity in the microbiome and higher levels of active Vitamin D (Thomas, et al., 2020). The researchers analysed both stool and serum samples of 567 older men, and then examined RNA sequences to identify the types of bacteria in the stool alongside the best-known quantifying measure of Vitamin D metabolites, LC-MSMS. Certain types of bacteria appeared more often in the microbiomes of those with more active Vitamin D. The majority of these types of bacteria were found to produce large quantities of butyrate. As is well known, lower levels of butyrate have been found in chronic conditions such as IBS. As such, it is seen as an important protector of the intestinal wall. Even more significantly, recent research has demonstrated that it plays a vital part as a modulator of optimal immune function. The problem now presented is intriguing, as these samples were collected from a wide range of people across different parts of the United States, and it would appear that it may demonstrate that it is not simply a case of how much Vitamin D a person can make, but how efficiently it makes active Vitamin D. Although largely metabolised by the liver and kidneys into its active form, it may well be that some residual metabolising occurs in the gut. Whilst this is somewhat speculation, it is certainly true to say that active Vitamin D has a symbiotic role in good microbiome health. As such, it has been found to modulate intestinal microbiome function and control antimicrobial peptide expression. As I have demonstrated, it is essential for good sleep and modifying inflammatory processes in acquired brain injury. Thus linking perfectly with more recent research around the gut biome itself and its potential for anti-

DOI: 10.4324/9780429199066-13

inflammatory effects in chronic conditions. The biome effects sleep disorders, especially but not limited to circadian problems, and it is therefore my third foundation of complete sleep management.

Background to the biome

More recent evidence indicates that microorganisms in the gut biome and the circadian genes interact with one another. Furthermore, sleep and healthy and unhealthy neurological states are regulated by the gut-brain axis. First, a few sentences on what we mean by the gut biome. The gut microbiome is all the micro-organisms, protozoa, fungi, viruses, and bacteria, and all their associated material that are present in the gastrointestinal tract. In other words, it contains neutral, negative, and positive microorganisms. It has been demonstrated to be crucial for a number of wide-ranging functions, including its effect on metabolism, immune responses and neuroendocrine processes. It is essential for nutrient, mineral, vitamin, and amino acid absorption and synthesis. Not only this, its fermented waste products, both indirectly and directly, are also critical for a similar range of functions, most obviously demonstrated through propionate and butyrate. To gain some understanding of the size and variety of the gut biome, the human body expresses some 20,000 eukaryotic genes, while the gut expresses 3.3 million prokaryotic genes. Estimates vary as to the totality of microbial cells, but best in the approximate range between 30 trillion and 400 trillion microbial cells, primarily, these are bacterial. As such, one suggestion was that, taken as a whole, our body is more bacteria than any other kind of cell. This is not entirely true and from what we understand, it is more likely to be almost a 1:1 ratio. Whatever the final calculation may be, it is fairly obvious that cells of such magnitude would have an enormous effect on all the other cells and functions of our body. The microbiome has been demonstrated to affect a number of different systems and functions within the human body. This can range from something as simple as Bifidobacteria being essential for the digestion of healthy sugars in breast milk, through to the ability of others to digest fibres which may prevent diabetes, heart disease, and even some cancers, as well as, the previously cited functions around brain health and immune function.

In the years to come, no doubt, extremely exciting developments will occur in our understanding about the function, dysfunction, and potential treatments for the other parts of the microbiome, perhaps even viruses. My own view is that once we have grown further in our understanding of the bacterial part of the biome, we will move towards understanding the power of the fungal component within the microbiome, as I am sure it will be as exciting as the discoveries that we are making for bacteria in the biome. Then perhaps the most difficult to understand the viral part of the biome will be tackled. These are exciting times for the gastrointestinal tract, even more exciting for those investigators looking at it.

One of the key people involved in research around bacteria in the gut and the person that coined the term 'probiotics' is the Russian scientist, Elie Metchnikoff (1845–1916). Whilst working at the Pasteur Institute in Paris, he was intrigued by the apparent benign effects of some bacteria he was studying. This became even more pronounced when he went to Bulgaria and found that the population where he was studying lived much longer than others. He began studying those Bulgarians who lived over 100 years of age. He attempted to find commonalities between such longevity and their life-style. He discovered that villagers in the Caucasus mountains drank on a regular basis a fermented yoghurt drink and those that lived the very longest drank this at least once a day. After analysis, he found that there was a particular lactobacillus in this drink, which he hypothesised to be key in their longevity. He won the Nobel prize in 1908 for his work on immunity and he is widely considered to be a father of immune research. He developed further theories around ageing and was a very strong advocate for the idea that ageing is actually caused by a build-up of toxic bacteria in the gut. Furthermore, that most bacteria emanating from lactic acids could prolong life, and as such he drank a version of the peasant drink of sour milk every day. From his pioneering research, many more probiotics have been discovered and the rise in their application to human conditions has widened. They may even be seen as the new-age panacea for health problems as mindfulness is for psychological problems.

In this regard, he could be said to have identified the three primary potential treatments for a disordered biome. This would be first probiotics, so-called because they are either benign or offer some level of health benefits, second prebiotics (e.g. inulin), which could be described as a form of compost for the good bacteria. Better and more diverse biomes can occur without additional probiotics supplementation, simply by increasing prebiotics. The third, and perhaps best way of increasing the essential diversity in the biome is to optimise the diet. In this sense, the by now famous blue zones of: Sardinia, Ikaria, Nicoya, Loma Linda, and Okinawa, together with their respective diets. These are in essence, although largely vegetarian, contain a couple of days' worth of meat or fish protein, but it is usually small and of extremely good quality.

Possibly the most well-known and rather obvious way in which our microbiome affects us is through disorders such as ulcerative colitis, irritable bowel syndrome (IBS), and the more general problems associated with 'leaky gut.' In particular over the early part of the 21st century, research on probiotics alleviating various health problems has taken a much more rigorous turn. I shall turn to these to allow the reader to see the background of research, its effectiveness and most pertinently, the key throughout all this research, as the essential boundaried nature of healthy biomes and healthy blood brain barriers. In regard to ulcerative colitis, the most recent systematic review and meta-analysis undertaken by Dang, et al. (2020), concluded that there were no statistical differences between the therapeutic effects of faecal microbiota transplantation and that of the commercially available probiotic, VSL#3. VSL#3 contains 450 billion bacteria from eight different bacterial strains per dose, and it has become one of the most widely used probiotics in research studies, particularly in bowel problems. Having said this, a wide range of studies, including many randomised controlled trials, now support probiotic usage in the IBS population, and several have demonstrated complete remission in all of the symptoms present, whilst others have at the very least shown a reduction in requirements for standardised medication. This latter point is important as the first line of treatment for such flare ups is often very strong steroids, including direct application of steroidal foam to the intestine (for recent meta-analyses, see for example McFarland and Dublin, 2008; Mari, et al., 2020).

Similar effectiveness has been found for the use of probiotics and the prevention of atopic dermatitis in children (see for example, Avershina, et al., 2016).

There was an inconsistency and perhaps even a controversy around probiotics and their usage for depression, however, more recently, studies are beginning to emerge that are much more supportive. To begin with, clear effects were found in rodent studies, when certain probiotics strains were provided to depressed mice, and these were pitted against Fluoxetine (SSRI), near exactly the same level of positive response was found in both conditions, leading those undertaking the studies to conclude that the right kind of probiotics are as effective as the standard SSRI. Increasingly, this has also been demonstrated in human studies of depression. A recent systematic review by Huang, Wang and Hu (2016) only utilised randomised control trials to identify the main effects of research. The meta-analysis showed that probiotics significantly reduced major depression disorder scores, it was particularly pronounced in the population under 60 years of age and their final conclusion was that 'we found that probiotics were associated with a significant reduction in depression, underscoring the need for additional research on this potential preventive strategy for depression'. A more recent review undertaken by a separate group of researchers in Rhode Island, Liu, Walsh and Sheehan (2019), concluded, in largely similar ways to the earlier meta-analysis. At this point they were able to study many more randomised controlled trials; indeed 34 were included in the analysis, and they were able to examine specific prebiotics in addition to probiotics. What they found was that whilst prebiotics did not differ significantly from placebo for depression, probiotic administration yielded small but significant reductions in both depression and anxiety. This effect was greater for the more severe clinical cases, in contrast to the milder ones studied in community settings. Once again, they end this paper by strongly asserting the need for much larger randomised controlled trials. At this point in the history of probiotic usage, from animal through to human clinical trials, it is undoubtedly the case that probiotics are of benefit. The question is which probiotics are of benefit to which specific clinical group at which particular time-frame. In this regard, having looked at a number of the studies, there is a similarity between this and the research around Vitamin D. It feels that we are at a point where we have some level of proof of concept, but we are yet to properly drill down into timing, dosage, and strain type for greatest effect. As a minimum, gut health and the biome should be viewed as a potent global health strategy. As such focus on the correct kind of diet and appropriate supplementation, in order to mitigate the possibility of going down with these types of illnesses in the first instance should be made. If depression and anxiety could be alleviated, to even a small degree, as an important co-morbid problem in brain injury and sleep disorders, would be eased. This in and of itself is reason alone for further thought.

At present, knowledge is, moving from the idea of simply providing a large quantity of different species of probiotics to a more considered approach based upon a more targeted species and one that is increasingly condition specific. A good example of such movement can be found in the systematic review undertaken by McFarland in 2020, which looked at strain and disease specificity. This view is more readily supported by the history of research around probiotics (see, for example, Ouwehand, et al., 2018), which is much more supportive of particular kinds of strains and particular kinds of cocktails in order to magnify the effect. As a quick aside, it may come to a point where a particular kind of combination of strains which would bear the most fruit. It would be unfortunate if we were to get this far in acknowledging the microbiome's importance, to come back to the erroneous assumption that there is one single magic bullet for any particular condition no matter how complicated.

Gut biome and sleep

There is growing evidence that the gut microbiome has significant influence over sleep quality. Early studies which examine sleep deprivation and the human microbiome yielded debatable results. To examine this more carefully, Smith, et al. (2019) looked at the relationship between sleep actigraphy measures and samples taken from the gut microbiome of participants. In addition, immune system biomarkers were measured to see what further relationships could be understood. The first and perhaps most unsurprising results were that, as with so many other research studies on the microbiome, the microbiome diversity was correlated with a wide range of positive sleep indicators. Thus, greater diversity was associated with increased sleep efficiency, total sleep time, and negatively associated with WASO. Moreover, positive associations were found between the total microbiome diversity and interleukin-6. In addition, they found the reverse, that lower total diversity was associated with poorer sleep measures and that, actually there were three specific bacterial taxa associated with the worst sleep quality. These were Lachnospiraceae, Corynebacterium, and Blautia. In a related review undertaken by Aguilar-López, et al. (2020), they made a strong case for the microbiome making a significant contribution to metabolism, circadian rhythm, and the immune system within an overarching theme of human homeostasis. An example of this is that intestinal bacteria display endogenous circadian rhythmicity and as such have direct effects on the intestinal circadian clock. There is a bidirectional aspect to this, in that overall circadian rhythmicity influences the composition of bacterial communities in the intestine. Also in 2020, Irwin, et al. published in the prestigious journal, *Nature*, another systematic review and meta-analysis, that examined studies from 1980 to 2019, and they concluded that supplementation significantly improved sleep quality.

One of the potential links with the gut biome and circadian rhythms is, once again, melatonin and vice versa, gut bacteria are sensitive to melatonin. Melatonin that is secreted into the gastrointestinal tract increases cultures of specific gut bacteria, for example E. aerogenes, however, it has no or little effects on K. Pneumoniae. Thus, some of the further entrainment mechanisms of melatonin actually have effects in the microbiome in specific ways (Paulouse, et al., 2016). This has been demonstrated almost by accident in studies of those patients undertaking a low Fodmap diet, which in addition to reducing inflammation in the gut itself and improving IBS, also has the unintended positive effect of improving sleep quality. As such, this is now being studied much more closely by Yan, et al. in a double blind randomised controlled trial, which started in 2020 in Australia. Animal and human studies have shown that alterations in prebiotic foods e.g. complex fibres such as inulin, in addition to probiotic supplementation e.g. VSL#3, have demonstrated improvements in sleep quality measures (Thompson, et al., 2020; Matenchuk, Mandhane and Kozyrskyj, 2020). The latter work was also interesting in their review of a significant number of papers in this area, which confirmed the gut microbiome and its diurnal circadian rhythmicity. It also noted that a number of papers found that persistent jet-lag, a high-fat, high-simple sugar i.e. a Western diet, impacted negatively on this circadian aspect through gut bacterial problems. They also went on to confirm that sleep fragmentation and short sleep duration were associated with gut dysbiosis and that at least some of these disturbances, which may also then have knock on effects on the hypothalamic-pituitary-adrenal axis (HPA), are predominantly mediated through overgrowth of harmful bacteria, a loss of diversity, and thus a relative dearth of 'good' bacteria. What is more, the end products of these 'negative' bacterial species which increasingly grow, induce fatigue. Part of the reason for the increase in fatigue is that the body is having to mop up things it shouldn't have to be mopping up, in inappropriate areas. In other words, it's not just a problem of 'leaky gut,' it is the direct immune response. Finally, they conclude that 'furthermore, probiotic supplementation has been found to improve subjective sleep quality.' I would further argue that it is not just self-perceived sleep quality that we are beginning to see objective evidence of, whether it is sleep actigraphy or full polysomnography (Sawada, et al., 2017; Higo-Yamamoto, et al., 2021; Harnett, et al., 2021). In many respects a great distance has been travelled, in some respects not that far from the powerful insights of the great Russian, Elie Metchnikoff. It is just that it took over 100 years to figure it out.

Gut biome and brain injury

An overgrowth of 'negative' bacteria has such profound effects on circadian rhythm and other parameters of sleep quality (both C and S), but they also have other wide-ranging negative effects on human beings. This next section could easily be summarised as: healthy gut equals a healthy immune system and a healthy brain. One of the ways that increasingly impoverished microbiomes behave is to increase gut permeability, allowing species to migrate into zones within the body that they should not be found in. This has a number of consequences, one of the most obvious of which is that bacterial products circulate and can potentially pass through the blood brain barrier. The circulation itself of these bacterial products causes localised and more general inflammatory responses from immune cells. A sophisticated paper on mast cell activation was produced by Girolamo, Coppola and Ribatti (2016). They discussed the dual nature of mast cell activation as on the one hand being potentially neuroprotective, but with increasing microbiome problems, chronic activation occurs leading to a potential link with neurodegenerative conditions (e.g. MS, Parkinson's). The absorption into the wider system of negative species, activates further inflammatory cascades in resident glial cells (i.e. microglia), which have been implicated in a wide range of chronic diseases, both neurological such as MS, but also in illnesses such as depression. Thus the biome and gut dysbiosis become important considerations. The concern with the early research in this area is that it was difficult to tease out the direction of causation. In other words, as the interlink between the enteric nervous system (ENS) and the central nervous systems (CNS) has been enhanced, the first thought was that these kinds of conditions would potentially lead one to have a more problematic microbiome. It has been estimated that the ENS contains somewhere between 250 and 500 million neurons. It is very similar to the brain in its neurotransmitter and functional nature, as such it has been described as the second brain. The ENS is increasingly understood as exceptionally complicated and sensitive. For example, it produces more than 30 different neurotransmitters including over 90% of the body's serotonin and has recently been discovered to have more neurons than the spine.

The ENS, is worthy of deep examination, but that is beyond the scope of this book. Suffice as to state, it is remarkably complex. It is influenced by the whole nervous system, yet can function completely independently from it, including the spinal cord and brain. It is innervated by two systems, primarily the vagus nerve but also the prevertebral ganglia. Yet, it can still function if the vagus nerve is severed. As well as the majority of the body's serotonin, over 50% of the body's dopamine and large quantities of GABA, galanin, and acetylcholine are made in the ENS. Moreover, most of these are made by the biome: specifically, bacteria. In many respects the ENS drives the CNS, not the other way around, and not what we may think from everything we used to be taught. Indeed, a short turn from the great neuroscientist Damasio's (e.g. 1994 and 2010) contention of feelings driving consciousness, leads very easily to feelings being derived from the ENS, and prior to this the bacteria in the gastrointestinal tract. In a very profound sense: we truly are what we eat! The implications of this knowledge are explored later in one of the foundations for treatment.

Later research, offers us the tantalising prospect that it is more, at least in the beginning of these conditions, a case of a disturbed microbiome and all of the surrounding issues we have discussed to be the root of such conditions. Studies supporting these views stem from those undertaken in other cases of trauma, so when patients undergoing surgery were analysed for diversity of species in their microbiome pre-surgery and immediately post-surgery, significant reductions were found (e.g. Lederer, et al., 2017; Howard, et al., 2017; Wenzel, 2019). Further specific research has now confirmed this as a significant problem in all forms of brain injury. A systematic review undertaken by Pathare, et al. (2020), found that all studies noted similar changes in reductions in bacterial populations and diversity after a traumatic brain injury and that this largely occurs within the first 48 hours after the event. Another study undertaken by Xia, et al. (2020), looked at gut dysbiosis after stroke and developed a Stroke-Dysbiosis Index (SDI), which was devised to assess microbiome differences. The higher the score on the index, the greater the problem of reduced diversity. It was also used blind to discriminate stroke patients from controls in the study, for the validation of the tool. Using logistical regression, they demonstrated that the SDI was an independent predictor of severity of stroke and was also significantly correlated with patient outcomes. They concluded that the application of their index would assist clinically but also add to the quantitative evidence linking gut microbiota to stroke and recovery from stroke.

In another study by Nicholson, et al. (2020), using a pre-clinical TBI cortical impact model, they demonstrated that significant changes were apparent in the gut microbiome as early as two hours after the injury and that some changes in the microbiome persisted through the entire study period. They concluded that changes in the microbiome may represent a biomarker to stage TBI severity and potentially be a predictor of outcome. It is very likely that the bidirectional nature of the problems associated with disturbed brain function and disturbed microbiome function have, as we demonstrated, a number of potential explanations with no single one being supreme, but likely involve neural, biochemical, microbiotic, and endocrine routes. Hence the focus on all the foundations not one.

Given one of the central themes of this book, what is most likely is that these interactions will also affect the inflammatory responses that we have discussed in other settings and for other reasons. If the brain injury itself of whatever kind started the gut dysbiosis and propelled the decrease in diversity of the potential strains. It would be very likely that the gut microbiota, having an important responsibility for managing the inflammatory responses across the body. Indeed most authorities suggest, as an exemplar, it is 70% of our immune system (Jung, Hugot and Barreau, 2010), it will be unsurprising that the net effect of this is that the reduced diversity and sheer reduction in numbers would make it more difficult for the microbiome to then manage the inflammatory responses, as it would normally. In this regard, an interesting systematic review and systematic analysis undertaken by Du (2020) looked at whether enteral nutrition which was supplemented with probiotics could reduce mortality and the infection rates of patients with severe brain injury. A total of 39 randomised controlled trials were included that involved over 3,000 patients. Their findings were startling. There was a decreased risk of infection, alongside a decreased risk of 7-, 14-, and 28-day mortality. There was also a decreased risk of gastrointestinal complications. Moreover, it shortened the length of time in ICU and the overall length of time in hospital. Their conclusion could not have been clearer. They stated that: 'Enteral nutrition supplemented with probiotics effectively decreases the risk of mortality, gastrointestinal complications, and infection and shortens the stays in ICU; therefore, it should be extensively adopted to manage these given patients.' Of course, for our purposes, the key question must be: is there evidence of chronic inflammatory processes that potentially link with chronic microbiome changes. The related question is do microbiomes after brain injury recover or not?

Certainly, there is ample evidence that activated inflammatory cells lead to increases in pro-inflammatory problems and potentially wider spread systemic inflammation. Also, that potentially the reverse could occur as was demonstrated in these early days, with fine tuning of diets or probiotic delivery to ensure a reduction of inflammatory processes. Theoretical papers in brain research are now espousing such news in mainstream journals, where once it would have been considered somewhat suspect, or as I have been repeatedly been accused of: 'eccentric.' For example, Weaver (2021), in the front-rank journal, *Brain Research*, would probably support my own view that standardised medical anti-inflammatory treatments in the early days are largely or wholly ineffective. She goes on to argue that we should be looking at targeting the intestine to reduce overall systemic inflammation. This is also closely related to the argument that Zhu, et al. (2018) make.

What is more, there is a deal of evidence from animal models and theoretical viewpoints, which largely concur with the notion that alterations in the gut biome will be long-lasting, as potentially long-lasting as the earlier research I cited on raised levels of pro-inflammatory markers. They are as I have demonstrated intimately linked. In a recent study that supports this, undertaken by Urban and his colleagues in 2020, the faecal microbiome profiles of 22 moderate to severely injured TBI patients were examined years after the insult. Through standardised gene and metagenomic sequencing, they found that there were significant differences in the TBI patients' microbiome when compared to the control group, and that these differences persisted after group stratification analysis. Notably, specific species were produced, and these were predominantly Prevotella and our old friend, Bacteroides. In addition, Ruminococcaceae were relatively abundant. They also concluded that there were altered amino acid levels and modified inflammatory markers. They suggest that the shifts cause a perpetuating cycle of problematic microbiomes and increased inflammatory processes. Ruminococcaceae has been previously demonstrated to be associated with amino acid deficient microbiomes and have been linked to disruptions in autonomic nerves (e.g. the vagus nerve). In addition, they suggest that it promotes small intestinal bacterial overgrowth, (SIBO), which further contributes to gut dysbiosis. They argue that the microbiome of the chronic TBI patients was most reminiscent of other chronic inflammatory conditions such as obesity and other systemic disease, including rheumatoid arthritis. In addition, the analysis revealed a decrease in the bacterium, Sutterella, which has been previously linked to MS. As with the earlier discussion when I pointed to studies demonstrating significant ongoing pro-inflammatory problems some 17 years after a TBI, the power of this study, even though it was a relatively small sample that were analysed, is that, on average, those studied were 20 years post-TBI.

Recommendations for improving the biome to ameliorate both sleep and brain injury problems

Having now fleshed out the idea of a healthy gut meaning a healthy immune system and a healthy brain, I will now turn to more precise characteristics of how that may be achieved. Prior to recommending specific probiotic strains and prebiotic supplements, it is important to underline the vital importance of a: 'good quality, highly diverse microbiome through diet alone.' Indeed, in many respects, going back to the ideas proposed by Metchnikoff does not mean we have to simply drink large quantities of sour milk. However, it does mean that we could at least contemplate that idea, in addition to more palatable foods, which primarily are fermented. Indeed, pure organic naturally fermented foods such as kimchi are certainly the way forward. This should not be taken as a green light for alcohol, bread, or other of these fermented products. There is then an obvious difference between fermented and naturally fermented foodstuffs.

Each of these has great potential as particular probiotic strains naturally occur as part of the fermentation process. Naturally, organic, unrefined, fermented foods include such things as sauerkraut, which contain, amongst other probiotic strains, L.Plantarum and B.Bifidum. Critically, the former has been demonstrated to significantly increase serotonin and dopamine in the brain and have in some work been shown to assist not only mood, but sleep. The latter have been demonstrated to have either direct anti-inflammatory effects or, as a precursor to modifying some hormones and vitamins (e.g. B12), has demonstrated improvements in mood as well. Returning to L.Plantarum briefly, it can also be found in high quantities in the Korean food, kimchi. Both naturally derived live Greek yoghurt and kefir contain large amounts of B.Infantis, B.Bifidum, and a wide range of Lactobacilli. As a quick aside, one Lactobacillus, L. Acidophilus, has been demonstrated to reduce cholesterol and improve nutrient absorption (potentially mitigating leaky gut), and B.Infantis has been shown to increase the relaxation response, it is proposed, via the vagus nerve in rats.

Apart from the repeated theme of inflammation and addressing chronic inflammatory responses in both sleep and brain injury conditions, another central theme of the book has been the importance of melatonin levels and the optimisation of those. A study by Wong, et al. (2014) undertook a randomised controlled placebo study, is illustrative. This examined the potential hypothesis that melatonin levels were linked to probiotic supplementation and that it was this linkage that was partly responsible for the success of the so-called mega probiotic supplement, VSL#3 in ameliorating IBS; which it has done repeatedly for over a decade. Forty-two IBS patients were randomly assigned to receive VSL#3 or a placebo for six weeks. All participants completed bowel and psychological questionnaires, underwent rectal sensitivity testing and saliva melatonin assays. They concluded that the supplement improved symptoms of IBS and increased rectal pain thresholds. Furthermore, symptom improvement was highly correlated with melatonin expression. Finally, they suggested as a consequence of this study, it may be that melatonin, in addition to the direct effects of the probiotics, was partly responsible for the success of these interventions. This study fits with others which suggest that there is a clear relationship between certain Bifidobacteria, which increase blood serum levels of tryptophan, serotonin, and melatonin, in that order. In another study by Bravo, et al. (2011), they found that after ingestion of a particular Lactobacillus strain, L.Rhamnosus, over a long period of time, specific brain regions were either more or less activated by GABA. Most relevant to insomnia and frequently co-morbid conditions of anxiety and depression, it produced a reduced GABA expression in the frontal lobes and the amygdala, but concomitantly increased GABA expression in the hippocampi. I need the reader to take a step back for a moment and fully digest (ahem) the import of the last couple of sentences. In other words, one small little bug produces a number of significantly important effects in different brain regions, all of which have the potential for ameliorating some of the most significant and disabling conditions that I have tried to address in this book.

To understand how mainstream this research has become, a paper by Strandwitz, et al., which was published in 2019 in the journal *Nature Microbiology*, combined RNA sequencing examination of stool samples with fMRI imaging in patients with major depressive disorder. The latter, of course, is a problem that has implicated GABA expression over decades, amongst other neurotransmitters. They concluded that: '...relative abundance of faecal Bacteroides are negatively correlated with brain signatures associated with depression.' Indeed, one particular bacteria, Bacteroides ssp., produced large quantities of GABA.

This leads us on to think about how, although increasing generally the diversity of the gut biome is largely a good thing, there are of course certain species that one must be more wary of. A good example is L.Reuteri, which has a more general effect on neuronal excitability, but also, counter-intuitively perhaps, has an anti-pain effect.

There is a small but growing literature on interventions for amelioration of various direct problems associated with brain injury and also indirectly on inflammatory processes. One example is the randomised trial undertaken by Wan, et al. (2020), which looked at the effects of enteral nutrition with and without the three species of B.Longum, L.Bulgaricus, and E.Faecalis on various measures of inflammatory responses in severely traumatically brain damaged people. They found, in those given these additional probiotics, significant reductions in ET-1, CRP, and IL-6, and TNF-α, amongst others.

So, in conclusion, a generally benign shift in the diet that focuses on fermented and sour foods of good quality will help promote a good quality and diverse quantity of bacteria in the gut. In this regard, a highly diverse gut biome is, as we have demonstrated, a good thing for the human system, but it may be that specific probiotics would also be useful for specific conditions. Again, this is likely to be more than one single strain. Great care should be taken with a dietary approach in terms of increasing the amount of fermented and/or sour foods, so as not to overwhelm a disturbed, impoverished microbiome too quickly. Once again, remember the old maxim that the body does not like too rapid metabolic changes. If this is true for dietary changes, it is equally true for the introduction of probiotic species to the microbiome. Where once it was difficult to get hold of a probiotic with a few million species in, which were found to not survive the acid in our stomach, we are now confronted with a wide range of super, mega, whopping, hundreds of billions of potential strains, which not only have been demonstrated to have positive effects. People have been put off using them because they have caused stomach upsets and other related problems, but these are usually short lived.

It is also true to say that at a basic level, it is much better to take a probiotic supplement with or before food as this reduces the potential for abdominal distress, as well as increasing the potential for optimisation of the supplement. A basic recommendation would be to take this 30 minutes before a meal of either the morning or lunch time. In that regard, it is also not recommended to have these for the night-time. Part of the reason for this is that the effects are potentially, in the short term, disruptive to the gut and might cause discomfort, keeping the individual awake. However, another reason is that the chemical translation from gut bacteria through to final neurotransmitter or other downstream effects obviously takes longer to produce, thus from gut bacterium to tryptophan to serotonin to melatonin takes a number of hours and in some neurochemical relationships, days. This also points to the fact that if one is making these changes, one has to do them consistently for a minimum of four weeks. Indeed, to get the full effects of changes in the gut biome, dietary and potentially probiotic, change needs to be sustained over the course of two or three months. Then they need to be continued, in this sense it is unlike a treatment period. It is for life, *for* life, for good quality of life.

Finally, a brief few words on prebiotics: increases in these are important. In this regard, obviously there is a danger with conditions such as IBS and those people who require low fibre diets for other reasons. Nevertheless, it is important for all except those in whom it is contraindicated, but only in extremely gradual titration, especially from naturally sourced vegetables, such as Jerusalem artichokes. Caution should be applied in prebiotic supplementation of harsher varieties such as chicory root (for inulin), as this has the potential to produce much more extreme stomach distress. If one can tolerate some, it would unquestionably help the overall system and act as a compost for the probiotics that have been introduced. So much depends on your biome starting point and any gut problems (such as IBS-d). This very neatly comes onto the problem of a lack of fibre in the Western diet. Once again rather, than adding supplements dietary changes are far better.

We have, for many years, assumed that because we have a super abundance of high-fat, high-protein diets based around modern farming methods, which have allowed us to eat more beef, pork, and other meats than in any point in human history, that we were nutritionally well-equipped as a species. In actual fact, the reverse is true. The best diets, as have been studied for many years, revolve around a balance of extremely good quality, natural ingredients. As a general rule of thumb, for main meals, we should aim for four (4) vegetarian meals, two (2) fish meals with vegetables, and one meat meal with vegetables each week. This is in addition to the basic balanced meals at other times. As well as specific prebiotic deficiency, this would neatly address one of the other most important areas of nutritional deficiency that we suffer from. Indeed, it may be described as a pandemic – a lack of fibre. The average Western diet has approximately 10–15 grams of fibre a day and we should be aiming for, depending on the source of information and the gender of the person, about 25–40 grams a day (there is a difference for the requirements of women and men, the lower figure is for women).

This focus on probiotics and the importance of these specific dietary requirements is now easy to make as there is an established literature on their use in a wide variety of conditions and a growing literature on their use in the specific conditions, we have discussed in regards to sleep, brain injury and inflammatory responses. This should not negate the fact that a general principle of good dietary recommendations should be followed. There is a large but contested literature around other dietary mechanisms that may be of benefit, most obviously these would include such commonplace dietary recommendations as Docosahexaenoic acid (DHA), one of the three components in Omega-3 fats, polyphenols for example blueberries, celery, certain lettuces, curcumin, and most especially other supposed antioxidant foods such as green tea. What is of most interest to me, having read and then re-read this literature, is that so much of the potential for the beneficial effects are much more likely to be found in plants, with some in fish but less in meat. Having said this there are some startling studies in the literature on dementia. Three stand out. The first was by Zang et al. (2017), who undertook a randomised controlled double-blind trial on a large group of older adults with mild cognitive impairments, using a DHA supplement or placebo. Two things stand out. First, it was done over 12 months, second, they measured hippocampal volume as a part of a multi-measure outcome approach. Significant increases in hippocampal volume were found in the DHA group and all other functional measures. The second study was a randomised double blind placebo controlled trail, which looked at folate, and folate plus DHA supplementation. After 0, 6, and 12 months, amyloid-β peptide markers were remeasured, in addition to using functional cognitive measures. All parameters demonstrated significant improvements in the folate plus DHA group (Bai, et al., 2021). These studies fit with the most recent review. In crude summary, the authors concluded that multi-modal approaches, which have nutrition and DHA through natural methods (e.g. oily fish) at their heart work reasonably well. They make one important caveat here. The earlier they are started the better the effect (Issek and Suchan, 2021). Even more strongly this is echoed by an earlier review, that stated: 'Conclusively, the interventions of the studies reviewed seem promising for individuals at risk of dementia, but not for those who are already diagnosed with dementia.' (Gkotzamanis and Panagiotakos, 2020). How does this equate to studies in acquired brain injury? Its basically that tentative support can be found for omega 3 and polyunsaturated fats (PUFAS) after any brain injury, largely due to their supposed anti-inflammatory effects (for an up to date review, see: Miao, et al., 2021). However, it seems, as with so many things discussed in this book the whole is better than a part. In that diets with whole fish and other natural sources of these fats, seem to have far better support than supplements. I return and will return again to the blue zone diets noted. It is the overall healthiness (or otherwise) of the diet that is more important than a supplement, derived from a part of a vegetable or fish segment.

I am not advocating veganism, we are not designed for this, I am advocating for a better balanced diet, which focuses more on vegetables than we have. Primarily, this is because, I'm obsessed with generating a good, diverse biome. In other words, I'm advocating the reverse of our Western dietary focus.

Chapter 14

Fourth foundation

Transition from ENS to the CNS

The following may be described primarily as the fourth foundation for insomnia, somewhat for hypersomnia but not at all for sleep apnoea. Briefly unpacking that latter clause, the research around vagal nerve stimulation whilst growing year on year, especially over the last ten to fifteen years, is very clear that it is contraindicated in those with sleep apnoea of central or obstructive origins. It is also very important to distinguish between the different types available. By this I mean the invasive technologies are undoubtedly the least safe. However, auricular forms are considered vastly safer and the completely self-instigated form that utilizes the body through chanting, and singing, is considered entirely benign. This of course underlines the first part of this chapter in my assertion of the paramount importance of diagnosis for those people with sleep problems and acquired brain injury. For if the proper diagnosis is made, then those with any form of sleep-disordered breathing (including apnoea) would be offered the gold standard treatments currently available.

First, it should be noted that the results from vagal nerve stimulation on sleep are not entirely benign. For many years vagal nerve stimulation was reported as a potential cause of central apnoeas and other sleep-related breathing problems in epileptic patients. In fact, respiratory dysfunction is the most frequently cited problem even though vagal nerve stimulation is largely well tolerated. As a consequence, the level and other parameters of the stimulation provided to manage the epilepsy needs careful titration to produce sometimes a compromise between breathing dysfunctions and improvements in epilepsy management.

Moving on from these caveats, the gut-brain axis is modulated by a number of factors, one of the most prominent being the large cranial nerve, called the vagus nerve. Famously, a German physiologist named Otto Loewi discovered in 1921, that through stimulating the vagus nerve he reduced heart rate. He went on to suggest that this was due to the release of a substance called,

DOI: 10.4324/9780429199066-14

"vagusstof." Vagusstof was later found to be the first neurotransmitter ever identified by scientists, now known as acetylcholine.

The nerve is so called from the Latin, Vagus, meaning to wander or ramble, as it traipses around many different parts of the body. Historically it was originally known as the pneumogastric nerve and is the tenth of the twelve cranial nerves. It is the main component of the parasympathetic nervous system (one of the main subsystems of the autonomic nervous system, the other being the sympathetic) and is critical for heart, lung, and digestive tract function. It originates in the brainstem (medulla oblongata), moves down through the neck into the thorax and onto the abdomen. In the neck, it provides innervation to the muscles of the pharynx and larynx, which are responsible for swallowing and vocalization. In the thorax it is critical for heart function, one of those aspects is to induce a reduction in heart rate. In the intestines, it supports peristaltic movements and secretions from various glands. Upon its return, it conveys important information to the locus coeruleus, medulla, the amygdala, and critically, given the previous descriptions in different chapters, the thalamus.

Activation of it can lead to a range of hormonal and neuropeptide changes; it is extremely important for the release of acetylcholine. This neurotransmitter is essential for movement, working memory, and various aspects of learning. Reductions in acetylcholine have been demonstrated to disrupt new learning and memory in several ways. Primarily, but not solely, due to its effects in the hippocampus (prompt for earlier section on memory!) One of the other unusual aspects of the vagus nerve, unlike so many other nerves in the body, is its strange capacity for regeneration. Even after vagotomy in a wide range of different species vagus afferents recover at over twice the rate of other damaged nerves. Although it has between 10% and 20% output (efferent functions), it is largely an afferent organ, conveying sensory information about a wide range of organs to the brain. Perhaps the most well-known treatment modality that utilizes the vagus nerve is that of vagal nerve stimulation for seizures. For almost a quarter of a century, this form of specific neurostimulation has helped control drug-resistant seizures in epileptic patients with good success. More recently, it has also been shown to help in drug-resistant cases of clinical depression (numerous, but see Bottomley et al., 2020, for an up-to-date review).

As knowledge has deepened in regard to the gut biome, various forms of vagal nerve therapy, including lower-level non-invasive stimulation, have been utilized increasingly for improving immune function and attempting to treat specific neurological disorders. Given the central themes of the book and the resultant treatments suggested, it may not surprise the reader that my own view is that this is the fourth foundation of sleep management treatments in regard to acquired brain injury. The so-called gut-brain axis is a complex system that has been shown to be both bidirectional in nature and has wide-ranging effects on the endocrine, immune and cognitive systems. Within this realm, we now describe as the ENS.

Critically, for our discussion, an important number of hormones and neuropeptides that the ENS produce into the circulatory system cross the blood-brain barrier. Once this was understood in general, the excitement of potential treatments for specific disorders mounted. During the last ten years in particular advances in understanding how potentially the gut microbiome modulates particular brain functions has rapidly developed, primarily using animal models, but latterly serious trials in human subjects, as was shown in the last part, have begun to emerge for brain disorders. The disorders attempted to be modified have been wide-ranging and are now increasing. They include autism, Alzheimer's disease, multiple sclerosis, Parkinson's disease, and even more recently nonprogressive acquired brain injuries such as stroke and TBI.

As well as the functions already described the ENS is intimately linked to both the innate and adaptive immune systems. Indeed, it is, alongside its partner, the brain, responsible for most of their functions and activities. Vagus nerve afferent pathways are linked to the critical hypothalamic-pituitary-adrenal (HPA axis). This of course amongst many other things coordinates the adaptive response to external and internal stressors. One such result could be increased systemic pro-inflammatory cytokines activated from the hypothalamus which ultimately leads to cortisol release from the adrenals. Furthermore, the vagus nerve is a central component of the neuroendocrine-immune system (axis). This axis coordinates behavioural, neural, and endocrine responses. This represents an important feature of inflammatory processing. For example, a pathogen which confronts the intestines necessitates the cytokine TNF-α derived from macrophages and dendrites coordinated by the vagus nerve. The dorsal motor nucleus of the vagus nerve responds in turn to the increased circulating TNF-α by altering motor activity. Three different anti-inflammatory pathways have been suggested for the vagus nerve. The first has been described and is the HPA axis, the second is something called the splenic sympathetic anti-inflammatory pathway, where once again the vagal nerve stimulates the splenic nerve to modify noradrenaline and acetylcholine. The acetylcholine then acts as an inhibitory neurotransmitter for TNF-α. The final pathway is known as the cholinergic anti-inflammatory pathway (CAIP), which once again sees the vagus nerve modify acetylcholine together with macrophages which result in further inhibition of TNF-α. This latter pathway is the most recent to be found and is not yet fully understood, however what we do know is that it is critical for the intestinal homeostasis and immune response. Thus, in summary, the inflammation sensing and activating and the inflammation suppressing roles outlined above are essential components of the vagus nerve and its related systems.

Therapeutic vagal nerve stimulation theory and research

During this section, vagal nerve stimulation techniques will be divided into three broad forms. First, a very brief introduction to implanted vagal nerve stimulation undertaken surgically. Second, transcutaneous auricular vagal nerve stimulation; which in essence is derived from methods and techniques from Chinese medicine in particular around acupuncture and the West's technological expertise (as was) in various electrical stimulation procedures. Third and finally the least invasive techniques derived from vagal nerve therapies, yoga, and voice training techniques, used in a wide range of mental health therapies (e.g. compassion focused therapy).

Throughout the early 20th century up until the late 1970s and 1980s, a wide range of experimental procedures were performed to understand the vagus nerve more completely. Latterly, procedures moved from animal models to human models and the first experimental vagal nerve stimulator was put forward in 1988 in America and closely patterned on the successful cardiac pacemakers fitted for those with electrical problems of the heart. In this case, the generator which is very small, perhaps the size of a small matchbox is surgically implanted below the collarbone. The wiring runs up the vagus nerve which is near the carotid artery. Once implanted this device sends out extremely low-level electrical impulses on a regular basis to stimulate the vagus nerve. It is on the left side of the body and thus stimulates the left vagus nerve; as the right has significant cardiac effects. This particular technique is not without its problems outside surgery itself and most commonly can produce cough, shortness of breath, and hoarseness. In addition, a significant portion of patients in large randomized control trials demonstrated an increase in seizure activity. However, these results were more negative in the early days of the pioneering work around this. As further understanding has been gained, subtle titration of the amplitude and frequency have occurred alongside improved delivery methods, resulting in the side effects, particularly the more serious ones, being reduced.

Subsequently, the FDA first approved treatments for intractable epilepsy in 1997. These procedures have now been rolled out across the world. Pioneering research revealed that the five primary metabolites found from vagal nerve-stimulated patients were acetylcholine, serotonin, dopamine, GABA, and ethanolamine. Specific forms of seizure activity were reduced and in particular, hippocampal changes in neurotransmitter release and concentration were found to have positive correlations with anticonvulsant effects.

More recent research has demonstrated that one of the causes of epilepsy is neuroinflammation, thus the role of vagal nerve stimulation in inflammatory pathway control has become of greater interest from a different theoretical perspective. At this point, research remains exciting but tentative but does conform to some of the previously discussed aspects in this book around the role of chronic neuroinflammation. If this was then tied together with the other conditions described that have relationships with inflammatory processes, we can further see the potential for vagal nerve stimulation on a much broader basis. An example is that more recently vagal nerve stimulation has been developed for treatment for refractory depression and some of these studies have demonstrated astonishing patient outcomes. For example, Bajbouj et al. (2010) found that two-year outcomes for vagal nerve stimulation showed that the majority of patients had equal to or better than 50% reduction in Hamilton Depression Scores from baseline.

Utilizing the same techniques applied for patients with migraines and the incredibly debilitating condition known as cluster headaches, showed pain reductions equal to or better than 56% immediately and 65% after two hours. This is especially heart-warming as cluster headaches are notoriously difficult to manage from a patient and clinician perspective; the pain frequently described as the most excruciating encountered.

Bringing this form of intervention up to date, and of most relevance to the current discussion, vagal nerve stimulation has shown success in other inflammatory diseases such as rheumatoid arthritis, cardiovascular diseases, and diabetes. Further studies are now being undertaken in the world of chronic inflammatory diseases. I shall now turn to vagal nerve stimulation and sleep and then vagal nerve stimulation and acquired brain injury specifically.

Transcutaneous auricular vagal nerve stimulation

Undoubtedly, part of the impetus for looking to low-level electronic therapeutic devices was partly to do with the ever-advancing sophistication and miniaturization in such, which enabled previously untransportable and unmanageable machines to be potentially at the behest of each patient. Not only this, but medication errors and side effects are undoubtedly the main variants which explains adverse events in the totality of medicine. Thus, vagal nerve stimulation or other cranial electrotherapy stimulation with extremely low levels of non-invasive micro-currents applied transcutaneously could be of great potential. The FDA in America was one of the first corporate bodies to allow clearance of such devices to be able to be purchased over the counter. An initial review undertaken by Kirsch (2002) looking at the previous 50 years of research on such stimulation for anxiety, found that 89% of the studies reported positive outcomes and that these studies were in mainstream peer-reviewed journals. The majority were double-blind placebo-controlled ones undertaken at prestigious American universities. Finally, and perhaps most interestingly, is that follow ups whether they were after a week or two years demonstrated some residual positive effects. Most modern transcutaneous devices stimulate the vagus nerve through the ear hence the frequently cited term in the literature, transcutaneous auricular vagal nerve stimulation (taVNS). However, it may also confusingly be known as tVNS or aVNS, depending on the author. In addition, it may be referred to as vagal and vagus nerve stimulation, once again both are the same thing. This convenient and non-invasive approach does not require surgery but has been demonstrated to be effective in insomnia.

Whilst it was largely developed through successful treatments of depression, a bi-product in outcomes was that it improved those with more insomnia symptomatology, (Fang, Rong and Hong, 2016). The idea for electrically stimulating the aural branch of the vagus nerve (ABVN) is a relatively more recent progression of cervical VNS, consequently there is not the breadth and depth of data available for standard stimulation that was outlined by Kirsch (2002).

However, there is a definite sea change that has occurred over the last six or seven years within this area. A good review was undertaken by Butt et al. (2019), which explored the neuroanatomical basis of this therapy. Their conclusion after reviewing cadaver, fMRI, animal, acupuncture, and clinical survey studies was that two areas were most suitable for stimulation. They suggested it was reasonable to surmise the concha and inner tragus were the most suitable locations for taVNS. They stress that other sites might be helpful but that the sites need determining through further research. One of the proposed mechanisms of action which has been studied is that tVNS triggers melatonin secretion, (Li, Zhai and Rong, 2014). It has also been demonstrated to increase sleep duration and decrease WASO.

However, tVNS has also demonstrated changes in endogenous opioids (endorphins) which are increased in the CNS as well as various other neurotransmitters (including encephalins and substance p). One of the other interesting effects of this stimulation has been repeated demonstrations of increases in GABA. Moreover, this is where recent research becomes extremely interesting for the purposes of this book, tVNS has been demonstrated to modify both inflammatory responses and neuronal plasticity (Olufsson et al., 2015). This study demonstrated that the anti-inflammatory effects, specifically in regard to a reduction in TNF-α lasted over 24 hours from one single stimulation. Similar, sustained anti-inflammatory and pain-relieving effects were demonstrated by Kovacic et al. (2017). The pain-relieving effects lasted for treble the amount of intervention time i.e. the intervention time was three weeks, but the pain-related anti-inflammatory effects lasted for nine weeks. A good recent review paper on the many potential therapeutic interventions emanating from vagal nerve stimulation but primarily focusing on chronic inflammatory conditions was undertaken by Bonaz and Pellissier (2016). What is more, in regard to recovery and in particular to neuroplasticity, beginning with rat models, Engineer et al. (2011) were able to reverse pathological changes in the cortex associated with tinnitus. Thus vagal nerve stimulation could offer potential for recovery directly, in terms of increasing recovery via enhancing neuroplasticity, as well as through its indirect modification of neuro-inflammation.

Vagal nerve stimulation has now begun to be seriously trialled in both progressive e.g. Alzheimer's Disease, (Kaczmarczyk et al., 2017) and non-progressive acquired brain injuries. The latter has produced a small but growing literature around such stimulation in animal models and latterly humans for traumatic brain injury and stroke, (Smith et al., 2005). In stroke an interesting study by Capone et al. (2017), targeted tVNS alongside robotic rehabilitation which significantly improved upper limb function after ischaemic stroke. In an earlier review of the potential for tVNS, Porter et al. (2011) suggested that its use when paired with therapeutic movements increased the neuroplasticity necessary to improve cortical representations of movements in the treatment of such disorders after stroke. Zhao et al. (2019) used tVNS in treating post-stroke insomnia for a case report. This also utilized fMRI to attempt to look at the neural mechanisms underlying the intervention. The team hypothesized that regulation occurred in the default mode network, visual cortex, and limbic circuits to produce the significant clinically positive effects on reducing insomnia post-stroke that was witnessed. As an aside, this article has appeared in the recent journal called '*Brain Stimulation.*' for those interested. Similarly, in their wide-ranging review, Feng, Bowden and Kautz (2013), discuss non-invasive brain stimulation techniques widely and conclude that these lead to sustained behavioural and clinical gains in patients who have suffered stroke with a variety of stroke-related deficits.

Least invasive techniques around vagal nerve stimulation

In this regard, we move from direct or indirect electrical stimulation to the idea of vagus nerve tone. Vagal tone is synonymous with the activity of the vagus nerve but it is also more recently subsumed under the idea of vagal tone exercising. A good review was undertaken by Kok et al. (2013). This is demonstrated with low vagal tone which has been linked to levels of high inflammation (see for example, Thayer and Sternburg, 2006, Bibevski and Dunlap, 2011). Low tone is also associated with more negative moods, heart attacks, strokes, and depression, all of which have strong implications for high levels of inflammatory problems. As demonstrated the vagus nerve plays an important role in metabolic homeostasis and helps control immune function and pro-inflammatory responses via the inflammatory reflex. The latter of which controls the innate immune response and subsequent inflammatory processes during pathogenic invasion and tissue injury. Disrupted immune regulation results in constant pro-inflammatory cytokine activity and chronic poorly controlled inflammatory processes. During the early part of this century further understanding of the cholinergic signalling system, especially understanding surrounding the release of acetylcholine, suppression of local and serum pro-inflammatory cytokine levels were established. Subsequent research has confirmed the anti-inflammatory effects of vagal nerve stimulation but as importantly through more gentle approaches such as yoga, meditation, and various chanting and singing techniques. Via such methods, research has shown, vagal tone can be improved and thus anti-inflammatory effects achieved. Early on Vagustoff (acetylcholine) was shown to be released through deep breathing and prolonged exhalations, acting as it did like a gentle tranquilliser.

A large body of research has now demonstrated parallel results from the breathing techniques described earlier (as a part of the sleep routine in the management section). Vagal nerve tone breathing techniques produce similar physiological effects, based around (but not solely) GABA, and acetylcholine release, increased vagal tone, and anti-inflammatory effects. As I said in that section, breathing is at the heart of relaxation therapies and potentially at the heart of breathing is improved vagal tone. There is a wealth of research and abundant literature regarding various forms of contemplative therapy approaches, for example yoga, various meditative approaches, mindfulness, and other similar more Eastern-based traditions. Within each of these, some form of modified breathing is central, whereas other aspects are different. Finding specific breathing or chanting research is less common but nearly always supportive of the potential vagal linkage.

A number of studies have substantiated both clear links with various methods of such vagal tone enhancement and solid positive effects. A good recent example of such research was published by Rankhambe and Pande (2020), two groups were divided between an "Om" chanting group and a control group. Whilst no specific instructions were provided in the control; the Om group were trained by a yoga teacher to use Om chanting with a simple breathing technique for 20 minutes each day for six days a week, over a total treatment period of four weeks. Hamilton Anxiety measures were taken for all participants before and after the study. Those in the treatment group demonstrated significant reductions in Hamilton scores, whilst the control group scores were unchanged at the end of the study period. These results are typical of the literature available. A more sophisticated approach to understanding the ancient art of chanting and its profound effects on the physiology and psychology of the human being can be found in Kalyani et al. (2011), who explored the fMRI correlations with a chanting, "Om," group or the placebo group pronouncing "ssss" as well as a third group in rest. Their conclusion was that the Om chanting group showed limbic deactivation which builds on studies on vagal nerve stimulation in depression and epilepsy that found identical results. This further confirms that training for such techniques would be considered an essential starting point for reducing inflammatory conditions in the least invasive manner possible. Later work by Rao et al., (2018) further confirmed through fMRI research that deactivation of limbic brain regions occurred during Om chanting. Outputs were particularly reduced from the significant area in and around the amygdala. Other research in this field has supported yogic breathing principles in cardiopulmonary effects (reduced hypertension: Posadski, et al., 2014), improving general physical function (Bussing, et al., 2012), mental health (Eberth & Sedlmeier, 2012; Pascoe, et al., 2015), stress and depression (the so-called PREVENT trial authored by Kuyken, et al., 2015), age-related decline (Gothe and McAuley, 2015), and most pertinently for this book, wide-ranging effects on chronic inflammatory processes. Two reviews of the latter will illustrate the importance of these studies. First, Gerbarg, Gootjes, and Brown., (2014) in their review of the literature on the neuropsychology of well-being concluded that both auricular vagal nerve stimulation and repetitive yogic chanting stimulated the vagus nerve and the mechanism of action appeared to be increased GABA, which is the reason for the overall calming effect. The other was a paper in the highest-ranked journal in the world: *Nature*, by Borovikova et al. (2000). In this paper they made the clear cut case that: vagal nerve stimulation attenuates the systemic inflammatory response to endotoxin (lipopolysaccharide) in rats.' They argued for afferent vagus nerve fibres being activated by endotoxin (or cytokines) stimulate the HPA-axis anti-inflammatory response. In this study, they found acetylcholine significantly attenuated the release of cytokines (including TNF-α, (IL)-1β, IL-6, and IL-18). Direct vagal nerve stimulation during lethal doses inhibited TNF-α synthesis and prevented the development of shock.

Finally, the question turns to not whether vagal nerve stimulation is effective for a number of inflammatory conditions and insomnia, but is it effective for those who have suffered an acquired brain injury. The answer would be most definitely, but potentially the precise mechanisms are still open for debate. In an early work by Bansal et al. (2012), in an animal model of vagal nerve stimulation and its modulatory effects on pro-inflammatory processes found that ghrelin appeared to be released through the stimulation of acetylcholine from previous studies examining vagal nerve stimulation. Amongst other processes ghrelin has been shown to be an important anti-inflammatory, as such key modifier of pro-inflammatory cytokines. They observed that the highest ghrelin levels were measured in those groups which had been in the treatment groups of vagal nerve stimulation. The lowest levels were found in the untreated groups who had suffered a traumatic brain injury. Not only this, but the VNS groups increased serum ghrelin and decreased TNF-α levels. A work of the same year Krzyzaniak et al. (2012), found that vagal nerve stimulation decreases blood-brain barrier disruption. These animal studies along with the literature around human studies of inflammatory processes and vagal nerve stimulation techniques allowed Neren et al. (2016) in their review of pre-clinical models of VNS and TBI, to conclude that VNS enhanced motor and cognitive recovery, attenuated oedema and inflammation, reduced blood-brain barrier breakdowns, and finally conferred neuroprotective effects. This latter functional gain is especially important to consider, as from even more recent research we now know increased permeability of the blood-brain barrier is found in those who carry the ApoE4 allele, which accelerates Alzheimer's disease and age-related brain changes more generally. What we also know, as discussed earlier, is the increased permeability, found in those with Vitamin D deficiency (a recent study on increasing memory impairments and poorer functioning blood-brain barriers i.e. increased permeability, see Banks, et al., 2021). It was now time to examine properly and fully the effects of VNS procedures in clinical settings. An extremely important animal model paper by Dong and Feng, (2018) who found in a rat model of VNS after traumatic brain injury that it modified the up regulation of orexin-A receptor type 1 expression in the prefrontal cortex, which had significant implications for VNS in assisting wake-promoting in the TBI population. In other words, this and other related VNS studies demonstrate that such stimulation can improve insomnia problems after a TBI. It may in addition to other interventions provide a further novel tool for assisting wakefulness in hypersomnia patients who have suffered an acquired brain injury. It is this latter ability, that when properly tuned, or stimulated, it appears to work in the way the host needs it to, that is most exciting. In other words, it is a single intervention which can affect the same system in seemingly opposite directions and final outcomes. In other words, it reminds us of certain neurotransmitters (e.g. serotonin) utilized at different times for different needs. Moroever, it may utilize different ones for the different needs of the host. For a more up-to-date review on orexin signalling see Tang, et al. (2019).

To conclude, from my own perspective it almost feels as though alongside the other foundations I have written about, vagal nerve stimulation by whatever of the safer means I have outlined can tune the system of impaired brains and inflammatory processes to the requirements needed for optimal health. Given that after all the decades of use of SSRI's and some anti-epileptic medication we are still not fully able to understand their specific effects in modulating the processes they are prescribed for, it seems extremely odd that these techniques outlined are still being debated. This is especially the case when its effects are much clearer and its mechanisms of action are better understood. That it is relatively cheap, easy to administer and safe, makes the current situation even more of a mystery. If it is further combined with my other foundations as presented, sleep will be enhanced alongside better outcomes in brain injury – of any aetiology.

The practical application of vagal nerve stimulation

The next part will work, as it should be, from least to the most invasive recommendations for interventions below.

Practical breathing, chanting, and similar vibrations

Many quick, simple, and effective breathing techniques could potentially be recommended for stimulating the vagus nerve. The 4–7–8 technique has already been recommended as one. Others are freely available, and revolve around, in essence, three elements: breathing more slowly (on average we take around 15 breaths a minute, aim for around seven, per minute), breathing much more deeply, especially focusing on the breath being diaphragmatic or the belly area (not the ribs), and finally slowing the exhalation for longer than the inhalation. Many different ones can be found in books, videos, and online; a preferred method should be chosen, then practice and perfect. Ideally the frequency should be three times each day for around three to five minutes at a time. If one is missed, ensure it is the middle of the day one and never the morning or evening one.

In addition, a cold shower for three times each week of 20 to 30 seconds has been demonstrated to enhance anti-inflammatory effects through vagal nerve stimulation, which will have a synergistic effect on the breathing techniques (c.f. Walsh 2018, for review of 30-second showers, recommended for athletes).

Finally, singing, humming, gargling, and most especially Om chanting are critically important to do at least twice daily (morning and evening) for a minimum of three (3) to five (5) whole minutes at a time. Interestingly, gargling after teeth cleaning in the evening and morning alone has been shown to improve sleep and or waking, through its vibration of the vagus nerve, (Kharrazian, 2013). The effects are thus best achieved through simple daily rituals, which gradually build up tone and have synergistic effects with the other foundations. In other words, bring these techniques into a daily routine, that is repeated "religiously."

Practical auricular transcutaneous vagal nerve stimulation (tVNS)

In order, to undertake tVNS adequately, one first needs to purchase a simple TENS machine. Many examples of these can be found online for between 15 and 150 pounds. Instructions are provided with each machine, or standard protocols can be found in the literature. One such has been provided by Deuchars et al. (2017), they suggest these parameters:

> We found that stimulation of the tragus part of the ear, using specific stimulation parameters established in the laboratory (200 μs pulses at 30 Hz 10–50 mA, which was slightly below the level at which subjects could feel the stimulus) significantly decreased the low-frequency to high-frequency ratio (LF/HF) of the heart rate variability, indicative of an increase in parasympathetic activity.

Gentle increases are key, early sources suggested the frequency should be increased to the point of slight pain. However, in the standard protocols developed by Deuchars et al. (2017 and 2019), the pulses were in essence just turned down to the level that is ever so slightly at awareness, but not painful. They should then be allowed to continue for the treatment period or session. In addition, if the tragus is unsuitable, most others recommend the concha as an effective alternative as a part of the ear to stimulate. Once again numerous practical examples of this can be found in self-help books (e.g. Rosenberg, 2017), magazines (Bergland, 2019), and online (JOVE.com). The latter contains a laboratory demonstration of what will be needed to set it up and proceed thoughtfully. It is a video that has been uploaded by the University of Carolina's Department of biomedical engineering alongside the Department of Paediatrics. These standard, sensible, clear accounts are essential base information needed prior to embarking on such stimulation. Another final recommendation is to read the paper already cited previously, that of Kanuisas et al. (2019; parts I and II); which outlines technical but somewhat practical ways of administering this therapy. An even more practical approach can be found in Yap et al. (2020) review of these procedures. Overall, these different approaches should be thought of as a safety net for a formulation, which should be distilled down to the individual concerned. Of equal tailored importance is frequency and duration of administration, in some studies only five (5) minutes of an evening once a day was shown to have some effect on inflammatory processes and sleep parameters. In most, however, the average was 15 minutes, up to 30 minutes once, or at most twice a day. These duration and frequency parameters were adequate to produce significant effects in a wide range of conditions, including pain, depression and most importantly, in the context of this book: inflammatory markers and sleep problems (e.g. Yap et al., 2020; Lermann et al., 2016; Deuchars et al., 2019).

In even more recent work, it has been used in the treatment of acquired brain injury. In one study published in 2020, Hakon et al. used this stimulation for four (4) hours a day, for eight (8) weeks in very severe brain-injured patients and demonstrated clinically significant gains in the treated group. One of the best summary papers of interventions, with exact specifications of the parameters involved in different aspects of stroke rehabilitation can be usefully consulted and has been provided by Ma et al. (2019).

Invasive procedures

This section should be seen as a very unlikely potential for the future. Moreover, at the moment it is being used for conditions that can affect sleep, or are the result of brain injury but not directly treating them. From surgically implanted devices in the 1990s for epilepsy, through to similar devices for refractory depression, to now the most recent successes in cluster headaches and migraines, clinical progress is being made. In time, treatments for acquired brain injury, and sleep; insomnia and hypersomnia may also become reality, but as I have demonstrated potentially irrelevant given the effectiveness of the other two techniques.

In conclusion, I am confident that what has been presented substantiates enough evidence, alongside the other foundations previously described to effect real significant clinical changes for the better, without having to resort to surgery, even if it were available now.

Part V

A recapitulation

Chapter 15

Final thoughts

Sleep

Prior to writing this book, I had for many years, been convinced that sleep had the potential to heal. An equally strong conviction being my belief in sleep's ability to heal a person who had suffered an acquired brain injury. Throughout the course of this book, it is obvious that sleep does not just heal but that it is an indispensable part of our existence. Indeed, healing although crucial is only one of the many important functions of sleep; it is as necessary as food and water. Moreover, disturbances in sleep, which takes up a third of our whole life, will negatively affect our mental and physical health to such a degree that ultimately it may, destroy it.

Of the two important sleep systems: Process C and Process S, we potentially have more control than perhaps initially perceived. However, we have somehow managed to lose our way during our post-industrial existence. This latter word is important, for if we are to thrive, and not simply exist, we must rectify and control better those aspects that impact upon both sleep processes. If we are to take better individual and collective responsibility for our sleep and what impacts upon it, we would undoubtedly start to thrive again. In this regard our health would blossom both mentally and physically, we would optimize our lives and those of our loved ones. In this effort, focus on quality and quantity are both critical. As such 7–9 hours are quantitatively important for our sleep each night, within this period adequate cycling through the requisite NREM and REM stages is qualitatively essential.

I have attempted to emphasize how neurochemicals and neurobiological processes are inseparable from functional outputs and inputs. An example, being the interplay between excitatory and somnolent neurotransmitters, which "battle" with one another to win through to either render us awake or asleep. As I have also established, attempting to run over these with an express train of a single powerful neurotransmitter derivative (e.g. the effects of a benzodiazepine), will produce unfortunate unintended consequences for the beautifully balanced system inside the human brain. Usually, the stronger the

DOI: 10.4324/9780429199066-15

drug compelling us to sleep, the worse we feel the following day. Thus, it only solves one-half of the whole. Far better to find natural analogues or alter our waking behaviours; or through a combination of these attempt more subtle approaches to enhance both sleeping and waking. A good example would be the interplay of artificially produced over-the-counter melatonin, bright light at the right time of day, and the four foundations as described.

Brain injury

Turning to acquired brain injury again. It is helpful to look at such injuries through the lens of neurochemistry and neurobiology. These devastating and all too common problems display a great many similar neurochemical and neu-robiological disturbances, which are equally crucial for the proper functioning of sleep. Many could be chosen to exemplify this argument. One such being the excitatory neurotransmitter glutamate, another would be the switching neuro-transmitter Orexin, and finally at the somnolent end, GABA.

Furthermore, the impact of such injuries irrespective of their initial ae-tiology has been shown to have remarkably similar secondary and tertiary neurochemical and neurobiological disturbances. Indeed, at the exact point the very hungry brain is in need of most nutrition, it fails to achieve it. Through the damage that has occurred, the necessary oxygen and glucose supplies are impaired. Not only has this but the energy supply system switched, to the less efficient and ultimately harmful, anaerobic generator. Then more rapid deterioration in the systems occurs as reactive oxygen species and other waste product-related damage starts to build.

Similarities can be also found in the kinds of ways that the body reacts to the subsequent neuroinflammatory processes. Once again, these kinds of processes then lead, in many instances, to self-perpetuating further neuro-inflammation. The key, in this regard, is to change the focus from efforts to ameliorate some potentially helpful acute inflammatory processes. Indeed, some of these early neuroinflammatory processes are essential for recovery and neuronal regrowth. Instead, greater focus should be placed on the de-leterious effects of chronic inflammatory processes. For, unlike the earlier acute inflammation, chronic inflammation provides absolutely no benefit to the person; it just accelerates decline.

Sleep and brain injury, and their mutual exacerbating effects

To understand the final outcomes better it is important to dwell for a mo-ment on the negative consequences of the interaction between sleep pro-blems and acquired brain injury.

It is unsurprising given the commonalities described at the neurochemical and neurobiological level that one will further exacerbate the other. In other words, organic damage will cause a previously healthy and well-slept individual to commonly develop sleep problems and then these will be perpetuated by the neurochemical changes outlined and potentially further harm both the sleep problem and the recovery from the brain injury. However, sleep problems in and of themselves can lead to various kinds of brain injury. Most obviously, sleep apnoea is an independent risk factor for stroke. Another example is the increased likelihood of Alzheimer's disease and its rapidity of decline. By this I mean, the more severe the sleep problem the steeper the decline and vice versa. Not only that but as has been established, these inflammatory problems are not resolved in a matter of months as was previously thought. Research over the last ten years has clearly shown these problems go on for many years; at this time, we know for at least twenty. This kind of research more adequately allows us to potentially pinpoint why there is a higher incidence of neurodegenerative diseases in a number of people who have sustained an acquired brain injury.

The very areas within the limbic system and associated neuroanatomical regions are the ones more likely to be harmed through the neurochemical and neurobiological processes examined by sleep problems and acquired brain injury. Thus, a significant part of the presentation and its ongoing difficulties are undoubtedly due to the synergistic effects of both acquired brain injury and sleep problems. Two primary examples would be the discussion around heterotopic ossification and other related bone density problems and all aspects of memory. These were chosen as exemplars, but I could have just as easily picked any number of other functional outcomes associated with these problems to illustrate the points I am trying to make. The remarkable similarity of deficits associated with either sleep or an acquired brain injury, can in many respects be seen as interchangeable.

Waste disposal, the glymphatic, and meningeal lymphatic system

All of this brings us neatly to the problem of waste disposal, which is a problem encountered with both sleep disturbances and acquired brain injuries. During the course of each and every day the brain, as a very greedy organ, produces vast quantities of waste due to its vast need for nutrients. There is then an equivalent amount of mopping up; cleaning and general showering needed to get rid of such an accumulation. This primarily takes place at night. Initially, it was thought only during NREM, but now it has been substantiated final finishing and polishing occurs in REM. Exciting discoveries in 2013, suggested a system that was named the glymphatic system (a portmanteau name derived from glial and lymphatic). This was thought to do the most important and thorough cleaning. However, a system first proposed by Paulo Mascagni, the great Italian anatomist, in 1787 is now, after centuries of debate, agreed upon. That is now known as the meningeal lymphatic system. It appears this offers a "speed flush" of some waste products. Increasingly, issues with one or both systems are linked with brain diseases; both degenerative and non-progressive (Da Mesquita, Fu and Kipnis, 2018; Ding et al., 2021).

After any acquired brain injury, both these systems are needed even more than usual. However, the toxic landfill of waste these injures produces, overwhelm both systems. That is without the usual daily sludge that is ordinarily produced; simply by being awake. Thus, the brain-injured person constantly has to keep up with a cleaning regimen with which they can no longer manage. As a quick aside, when one thinks of CSF drainage problems, one's initial thoughts turn to signs. These often involve fatigue increases, which overwhelm the patient and their ability to function, eventually leading to drowsiness, sleep, and worse. Before oedema sets in perhaps the initial signs of increased fatigue are actually to do with the brain's inability to clean itself properly. Too much toxic waste, which it can no longer clean away properly.

The other critical factor of these waste disposal systems is that they are fully and completely intact. From what we now know, this is far from the case. In addition, the more pronounced the acquired brain injury the more likely these systems will be damaged. It would be hard enough for a fully functioning waste disposal system to work efficiently enough to remove such a tsunami of waste, let alone a damaged one. Specifically, it would appear that the influx of cleaning fluids arrives, but their efflux, or exits are impaired in various ways. There is still more to be discovered, but this alone can explain so much clinically, including aspects of fatigue.

What is more, highly important structures for sleep and brain functioning are differentially damaged through the neurochemical and neurobiological damage described: some more than others. Structures in the limbic system, so important for good sleep and wake functioning, such as the hypothalamus and hippocampus seem most sensitive to the waste products generated. This is not an exclusive argument but a more subtle one that recognizes other regional damage, being properly aware of the regenerative power of a healthy hippocampal region for recovery. The symbiotic relationship of sleep and acquired brain injury is further aggravated.

The solutions should tackle both sleep and brain injury, to allow the power of sleep to shine

Thought needs to be given to both brain injury recovery and sleep problems and their joint treatment. What I am suggesting does in no way mean existing neuro-rehabilitation efforts are without great merit. I am suggesting these efforts can be fine-tuned and further enhanced, through utilizing a slightly different vantage point. This is even more pertinent in the longer term as weeks turn to months, then years.

In the first instance, know the problem. Proper assessment and diagnosis are critical for success. If one of the three main sleep problems is identified then specific recommended treatments may be commenced. Thus, during the course of this book, the treatments for insomnia, hypersomnia, and sleep apnoea (the top three sleep problems in those who have suffered a brain injury) have been argued for. However, what is crucial for their full success is that each treatment should be done alongside the four foundations if clinically meaningful results are to be obtained. Indeed, some that have been previously cited have suggested that only one of these would be enough to ameliorate the sleep disorders mentioned. However, I feel this is overstated, and also does not adequately comprehend their interplay. It is the *linkage* between them and their related nature which makes the interventions combined so powerful.

In crude summary: herbal remedies and or melatonin supplements can be fully recommended for insomnia. They are as effective as the standard medical pharmaceutical approach, but importantly with far fewer short, medium and long-term side effects. They are less likely to invade the day and compromise wakefulness. And with even greater force, unlike the standard treatments, they all have important anti-inflammatory effects thus tackling both problems at once. For hypersomnia, it is the reverse; the standard medical pharmaceutical approach is actually underutilized and would be of great benefit. This is most especially true with the new H3 medication, for example, as it has the least harmful side effects. That is with the simple proviso, that strong blue light therapies have been attempted first, as they are powerful, underutilized, and will have the additional benefit of easing common co-morbidities, such as depression and anxiety. Finally, sleep apnoea needs treatment at a specialist clinic and the gold standard treatment is CPAP technologies. However, this is also a sleep problem, which is under-diagnosed and thus under-treated for the acquired brain-injured population.

The four foundations are critical for the success of each of these three approaches in order to optimize sleep and recovery from an acquired brain injury.

First, there are a number of key elements that are necessary for a thoroughgoing and rigorous sleep management approach (not a simplistic sleep hygiene, top tips, or "hack"). They should be applied with a deep understanding of the behavioural principles underlying them.

To begin with, exercise is fundamental. It is also the only of the waking elements of this approach that has been demonstrated to have positive effects on the glymphatic system. If this system's power to aid recovery has been understood properly, then exercise should simply be done each day, depending on the individual's level of fitness, time point since the brain injury, age, and other co-morbidities. That something is done is non-negotiable.

Second, true sleep opportunity time needs calculating.

Third, a great deal of emphasis needs to be placed on behavioural routines. Temperature of the body and the room one sleeps in, alongside a complete lack of noise, are the fourth and fifth elements. The sixth element may be argued to be the most important of all: that is the proper management of external light sources, both unnatural and natural. Finally, naps, and weekend lay-ins, need a proper understanding; in the context of an individual's detailed knowledge of their own sleep profile.

The second foundation relies on the hormone, vitamin D, which should be taken in a supplement form alongside the available natural production via sun exposure. This is critical except perhaps for a very very small period in the height of summer. In order to adequately understand the amount required during the year; proper testing needs to be done. The interesting aspect of this hormone is its potential ability to ameliorate all three of the primary sleep problems discussed, as well as providing support for reducing inflammatory processes that have become self-perpetuating, long after injuries occurred. Part of this is its ability to down-regulate pro-inflammatory cytokines. It has a vital role in neuronal and glial tissues in particular, through buffering calcium.

It has a similar pattern of receptors that are found to be essential in those regions for sleep regulation. Furthermore, it has been shown that deficiency causes reduced sleep duration, generally poorer sleep quality, and specific measures such as declines in sleep latency, and subjective reports of negative sleep quality. To this end, a number of state health boards are now recommending supplementation after testing for deficiencies (e.g. Eire, South Korea). In a short time, no doubt, as further evidence accumulates, it will, once again, be supplemented by all states across the globe.

The third foundation is the biome. The aim should be at a minimum to increase the diversity of the biome through a better diet. This diet should be based on one of the dietary blue zone ones. In addition, increased fermented foods will be needed to promote the right kind of overall healthy biome. Finally, certain probiotic supplements may in addition be added to achieve the necessary improvement. Samples of faeces can now be sent for analysis at registered legitimate laboratories, which can then recommend certain types of strains over others. This will become commonplace, as consumers realize the power of the biome for good or ill. As shown these strains can have such specific effects on particular areas of brain function, which are simply startling. One such is the effects of a probiotic strain that makes expression of GABA in the limbic system much higher. When tailored cocktails of strains become more widely available changes will be even more profound. In the meantime, basic changes to the diet, increases in fermented products, and some more benign probiotic supplementation will provide enormous benefits. It is necessary to know that a healthy gut will produce a healthy immune system, ultimately a healthier brain (even after a brain injury), and thus healthier sleep.

Finally, the fourth foundation is some form of vagus nerve intervention. This could be as basic as chanting, singing, or other specific physical exercises, through to transcutaneous auricular vagal nerve stimulation, ending with invasive vagus nerve stimulation, done through surgery. Unquestionably, the first form, as long as it is repeated a number of times each day, will work alongside the other three foundations. As an example, one of the effects of such exercise is to increase the appropriate release of the neurotransmitter acetylcholine. As shown, acetylcholine has a number of significant purposes. It is essential for movement, working memory, and many aspects of learning. Without adequate functioning of the acetylcholine system, REM sleep and wakefulness will not occur. Not only this but without acetylcholine there is no essential paralysis during REM sleep; it has been hypothesized that reduced acetylcholine during this stage leads to some of the acting out disorders seen during REM. From vagal nerve exercises through acetylcholine to a balance of specific wake and sleep processes.

Sleep and wake cycles are intimately linked, each needs to be optimized. One cannot be addressed without adequately understanding the other. Other positive effects on the CNS from the ENS via the vagus nerve have been explored; numerous results have also been demonstrated on the orexin system, amongst others. From probiotic strains, through their neurotransmitter manufacture, via the vagus nerve, to finally, better brain functioning.

A central consideration is the linkage of these interventions for two purposes. By that, the behavioural management (including internal and external environmental considerations), and the further three foundations (Vitamin D, the biome, and vagal nerve exercises); all are necessary to ameliorate sleep problems and promote brain injury recovery. Links begin with the need for light at the right time. This then produces a hormone, which acts upon a wide range of functions and responses, including sleep and immune systems, and thus important inflammatory processes. Then the biome, a powerhouse of bacterial neurotransmitter manufacturing, and the basis of around 70% of the body's immune system, in toto the ENS is harnessed for the benefit of sleep and brain injury recovery. It is not too strong to say that in this regard the ENS should be viewed as *more* important than the CNS. The link here between the ENS and the CNS can itself be utilized to produce powerful effects, but as importantly the biome and the vitamin D will be enhanced in their intervention through its healthier tone. The most important reason for recommending these foundations is the dual targeting of improved sleep, alongside the improvements in chronic inflammatory processes and thus better outcomes for those with an acquired brain injury. Then when one of the three specific treatments for a specific sleep disorder is chosen the final outcome will be even more powerful.

It is at this point that sleep will begin to heal again.

References

Adams, J.H., Graham, D.I. and Gennarelli, T.A., 1983. Head injury in man and experimental animals: neuropathology. In: *Trauma and Regeneration*. Vienna: Springer, pp. 15–30.

Adams Jr, H.P., Bendixen, B.H., Kappelle, L.J., Biller, J., Love, B.B., Gordon, D.L. and Marsh 3rd, E.E., 1993. Classification of subtype of acute ischemic stroke. Definitions for use in a multicenter clinical trial. TOAST. Trial of Org 10172 in Acute Stroke Treatment. *Stroke*, *24*(1), pp. 35–41.

Ageborg Morsing, J., Smith, M.G., Ögren, M., Thorsson, P., Pedersen, E., Forssén, J. and Persson Waye, K., 2018. Wind turbine noise and sleep: pilot studies on the influence of noise characteristics. *International Journal of Environmental Research and Public Health*, *15*(11), p. 2573.

Agargun, M.Y. and Cartwright, R., 2003. REM sleep, dream variables and suicidality in depressed patients. *Psychiatry Research*, *119*(1–2), pp. 33–39.

Ahmadi, M., Khalili, H., Abbasian, L. and Ghaeli, P., 2017. Effect of Valerian in preventing neuropsychiatric adverse effects of efavirenz in HIV-positive patients: a pilot randomized, placebo-controlled clinical trial. *Annals of Pharmacotherapy*, *51*(6), pp. 457–464.

Åkerstedt, T., Ghilotti, F., Grotta A., Zhao H., Adami, H.-O., Trolle-Lagerros Y. and Bellocco R., 2019. Sleep duration and mortality – Does weekend sleep matter? *Journal of Sleep Research*, *28*(1), e12712.

Albrecht, J.S. and Wickwire, E.M., 2020. Sleep disturbances among older adults following traumatic brain injury. *International Review of Psychiatry*, *32*(1), pp. 31–38.

Alexiev, F., Brill, A.K., Ott, S.R., Duss, S., Schmidt, M. and Bassetti, C.L., 2018. Sleep-disordered breathing and stroke: chicken or egg? *Journal of Thoracic Disease*, *10*(Suppl 34), p. S4244.

Alkozei, A., Smith, R., Pisner, D.A., Vanuk, J.R., Berryhill, S.M., Fridman, A., Shane, B.R., Knight, S.A. and Killgore, W.D., 2016. Exposure to blue light increases subsequent functional activation of the prefrontal cortex during performance of a working memory task. *Sleep*, *39*(9), pp. 1671–1680.

Alu, M., Kiselev, V.N. and Loboda, E.B., 1977. Cerebrospinal fluid melatonin in diseases of the nervous system. *Zhurnal nevropatologii i psikhiatrii imeni SS Korsakova (Moscow, Russia: 1952)*, *77*(12), pp. 1814–1816.

Ames 3rd, A., Wright, R.L., Kowada, M., Thurston, J.M. and Majno, G., 1968. Cerebral ischemia. II. The no-reflow phenomenon. *The American Journal of Pathology*, *52*(2), p. 437.

Andreotti, J.P., Silva, W.N., Costa, A.C., Picoli, C.C., Bitencourt, F.C., Coimbra-Campos, L.M., Resende, R.R., Magno, L.A., Romano-Silva, M.A., Mintz, A. and Birbrair, A., 2019, November. Neural stem cell niche heterogeneity. In *Seminars in Cell & Developmental Biology* (**Vol. 95**, pp. 42–53). Academic Press.

Anjum, I., Jaffery, S.S., Fayyaz, M., Samoo, Z. and Anjum, S., 2018. The role of vitamin D in brain health: a mini literature review. *Cureus*, *10*(7), e2960.

Antonenko, D., Diekelmann, S., Olsen, C., Born, J. and Mölle, M., 2013. Napping to renew learning capacity: enhanced encoding after stimulation of sleep slow oscillations. *European Journal of Neuroscience*, *37*(7), pp. 1142–1151

Antonino, D., Teixeira, A.L., Maia-Lopes, P.M., Souza, M.C., Sabino-Carvalho, J.L., Murray, A.R., Deuchars, J. and Vianna, L.C., 2017. Non-invasive vagus nerve stimulation acutely improves spontaneous cardiac baroreflex sensitivity in healthy young men: A randomized placebo-controlled trial. *Brain Stimulation*, *10*(5), pp. 875–881.

Antrobus, J.S. and Bertini, M. eds., 2013. *The neuropsychology of sleep and dreaming*. Psychology Press.

Arabi, S.M., Sedaghat, A., Ehsaei, M.R., Safarian, M., Ranjbar, G., Rezaee, H., Rezvani, R., Tabesh, H. and Norouzy, A., 2020. Efficacy of high-dose versus low-dose vitamin D supplementation on serum levels of inflammatory factors and mortality rate in severe traumatic brain injury patients: study protocol for a randomized placebo-controlled trial. *Trials*, *21*(1), pp. 1–8

Aserinsky, E. and Kleitman, N., 1953. Regularly occurring periods of eye motility, and concomitant phenomena, during sleep. *Science*, *118*(3062), pp. 273–274.

Aserinsky, E. and Kleitman, N., 1955. Two types of ocular motility occurring in sleep. *Journal of Applied Physiology*, *8*(1), pp. 1–10.

Ashina, H., Iljazi, A., Al-Khazali, H.M., Ashina, S., Jensen, R.H., Amin, F.M., Ashina, M. and Schytz, H.W., 2020. Persistent post-traumatic headache attributed to mild traumatic brain injury: Deep phenotyping and treatment patterns. *Cephalalgia*, *40*(6), pp. 554–564.

Atkinson, R.C. and Shiffrin, R.M., 1968. Human memory: A proposed system and its control processes. In *Psychology of Learning and Motivation* (**Vol. 2**, pp. 89–195). Academic Press.

Auld, F., Maschauer, E.L., Morrison, I., Skene, D.J. and Riha, R.L., 2017. Evidence for the efficacy of melatonin in the treatment of primary adult sleep disorders. *Sleep Medicine Reviews*, *34*, pp. 10–22.

Avershina, E., Rubio, R.C., Lundgård, K., Martinez, G.P., Collado, M.C., Storrø, O., Øien, T., Dotterud, C.K., Johnsen, R. and Rudi, K., 2017. Effect of probiotics in prevention of atopic dermatitis is dependent on the intrinsic microbiota at early infancy. *Journal of Allergy and Clinical Immunology*, *139*(4), pp. 1399–1402

Ayerbe, L., Ayis, S., Wolfe, C.D. and Rudd, A.G., 2013. Natural history, predictors and outcomes of depression after stroke: systematic review and meta-analysis. *The British Journal of Psychiatry*, *202*(1), pp. 14–21.

Baddeley, A., Eysenck, M.W. and Anderson, M.C., 2020, *Memory*, 3rd edn, Routledge.

Bai, Dong et al. 'Effects of Folic Acid Combined with DHA Supplementation on Cognitive Function and Amyloid-β-Related Biomarkers in Older Adults with Mild Cognitive Impairment by a Randomized, Double Blind, Placebo-Controlled Trial'. *J. Alzheimer's Disease*, *81*(1), pp. 155–167, 2021

Baird L.C., Newman C.B., Volk H., Svinth J.R., Conklin J., Levy M.L., 2010 Nov Mortality resulting from head injury in professional boxing. Neurosurgery. *67*(5), pp. E519–E520

Bajbouj, M., Merkl, A., Schlaepfer, T.E., Frick, C., Zobel, A., Maier, W., O'Keane, V., Corcoran, C., Adolfsson, R., Trimble, M. and Rau, H., 2010. Two-year outcome of vagus nerve stimulation in treatment-resistant depression. *Journal of Clinical Psychopharmacology*, *30*(3), pp. 273–281

Balachandran, J.S. and Patel, S.R., 2014. Obstructive sleep apnea. *Annals of Internal Medicine*, *161*(9), pp. ITC5-1.

Baran, Bengi, Janna, M. and Rebecca, MC Spencer, 2016. "Age-related changes in the sleep-dependent reorganization of declarative memories." *Journal of Cognitive Neuroscience 28*(6), pp. 792–802

Barnaś, M., Maskey-Warzęchowska, M., Bielicki, P., Kumor, M. and Chazan, R., 2017. Diurnal and nocturnal serum melatonin concentrations after treatment with continuous positive airway pressure in patients with obstructive sleep apnea. *Pol Arch Intern Med*, *127*(9), pp. 589–596.

Barnes, M., Bennet, E. and Etherington, J., 2018. Acquired brain injury and neurorehabilitation time for change: All party parliamentary group on acquired brain injury report. *UKABIF: UK*.

Barth, J.T., Freeman, J.R., Broshek, D.K. and Varney, R.N., 2001. Acceleration-deceleration sport-related concussion: the gravity of it all. *Journal of Athletic Training*, *36*(3), p. 253.

Basińska-Szafrańska, A., 2021. Metabolic diversity as a reason for unsuccessful detoxification from benzodiazepines: the rationale for serum BZD concentration monitoring. *European Journal of Clinical Pharmacology*, *77*, pp. 1–14

Bassetti, C., Aldrich, M.S., Chervin, R.D. and Quint, D., 1996. Sleep apnea in patients with transient ischemic attack and stroke: a prospective study of 59 patients. *Neurology*, *47*(5), pp. 1167–1173.

Bathel, A., Schweizer, L., Stude, P., Glaubitz, B., Wulms, N., Delice, S. and Schmidt-Wilcke, T., 2018. Increased thalamic glutamate/glutamine levels in migraineurs. *The Journal of Headache and Pain*, *19*(1), pp. 1–8.

Baumann, C.R. and Bassetti, C.L. 2005 Oct Hypocretins (orexins) and sleep-wake disorders. *Lancet Neurol.* 4 (10): 673–682

Baumann, C.R., Werth, E., Stocker, R., Ludwig, S. and Bassetti, C.L., 2007. Sleep–wake disturbances 6 months after traumatic brain injury: a prospective study. *Brain*, *130*(7), pp. 1873–1883

Baumann, C.R., 2012. Traumatic brain injury and disturbed sleep and wakefulness. *Neuromolecular Medicine*, *14*(3), pp. 205–212.

Baylan, S., Griffiths, S., Grant, N., Broomfield, N.M., Evans, J.J. and Gardani, M., 2020. Incidence and prevalence of post-stroke insomnia: a systematic review and meta-analysis. *Sleep Medicine Reviews*, *49*, p. 101–222.

Bedrosian, T.A. and Nelson, R.J., 2017. Timing of light exposure affects mood and brain circuits. *Translational Psychiatry*, *7*(1), pp. e1017–e1017

Beersma, D.G., 1998. Models of human sleep regulation. *Sleep Medicine Reviews*, *2*(1), pp. 31–43.

Bent, S., Padula, A., Moore, D., Patterson, M. and Mehling, W., 2006. Valerian for sleep: a systematic review and meta-analysis. *The American Journal of Medicine*, *119*(12), pp. 1005–1012.

Bergland, C., 2019. Longer Exhalations are an Easy Way to Hack your Vagus Nerve. *Psychology Today*, [online]

Bernhofer, E.I., Higgins, P.A., Daly, B.J., Burant, C.J. and Hornick, T.R., 2014. Hospital lighting and its association with sleep, mood and pain in medical in-patients. *Journal of Advanced Nursing*, *70*(5), pp. 1164–1173.

Bettcher, B.M., Wilheim, R., Rigby, T., Green, R., Miller, J.W., Racine, C.A., Yaffe, K., Miller, B.L. and Kramer, J.H., 2012. C-reactive protein is related to memory and medial temporal brain volume in older adults. *Brain, Behavior, and Immunity*, *26*(1), pp. 103–108.

Bibevski, S. and Dunlap, M.E., 2011. Evidence for impaired vagus nerve activity in heart failure. *Heart failure reviews*, *16*(2), pp. 129–135

Bigué, J.L., Duclos, C., Dumont, M., Paquet, J., Blais, H., Menon, D.K., Bernard, F. and Gosselin, N., 2020. Validity of actigraphy for nighttime sleep monitoring in hospitalized patients with traumatic injuries. *Journal of Clinical Sleep Medicine*, *16*(2), pp. 185–192.

Bloom, H.G., Ahmed, I., Alessi, C.A., Ancoli-Israel, S., Buysse, D.J., Kryger, M.H., Phillips, B.A., Thorpy, M.J., Vitiello, M.V. and Zee, P.C., 2009. Evidence-based recommendations for the assessment and management of sleep disorders in older persons. *Journal of the American Geriatrics Society*, *57*(5), pp. 761–789.

Blumbergs, P.C., Scott, G., Manavis, J., Wainwright, H., Simpson, D.A. and McLean, A.J., 1994. Stalning af amyloid percursor protein to study axonal da-mage in mild head Injury. *The Lancet*, *344*(8929), pp. 1055–1056.

Boden-Albala, B., Roberts, E.T., Bazil, C., Moon, Y., Elkind, M.S., Rundek, T., Paik, M.C. and Sacco, R.L., 2012. Daytime sleepiness and risk of stroke and vascular disease: findings from the Northern Manhattan Study (NOMAS). *Circulation: Cardiovascular Quality and Outcomes*, *5*(4), pp. 500–507.

Bodnar, C.N., Morganti, J.M. and Bachstetter, A.D., 2018. Depression following a traumatic brain injury: uncovering cytokine dysregulation as a pathogenic me-chanism. *Neural Regeneration Research*, *13*(10), p. 1693.

Bogdanov, S., Brookes, N., Epps, A., Naismith, S.L., Teng, A. and Lah, S., 2019. Sleep disturbance in children with moderate or severe traumatic brain injury compared with children with orthopedic injury. *The Journal of head trauma re-habilitation*, *34*(2), pp. 122–131.

Bogdanov, S., Naismith, S. and Lah, S., 2017. Sleep outcomes following sleep-hygiene-related interventions for individuals with traumatic brain injury: a sys-tematic review. *Brain injury*, *31*(4), pp. 422–433

Bolte, A.C., Dutta, A.B., Hurt, M.E., Smirnov, I., Kovacs, M.A., McKee, C.A., Ennerfelt, H.E., Shapiro, D., Nguyen, B.H., Frost, E.L. and Lammert, C.R., 2020. Meningeal lymphatic dysfunction exacerbates traumatic brain injury pathogenesis. *Nature communications*, *11*(1), pp. 1–18.

Boly, M., Massimini, M., Tsuchiya, N., Postle, B.R., Koch, C. and Tononi, G., 2017. Are the neural correlates of consciousness in the front or in the back of the cerebral cortex? Clinical and neuroimaging evidence. *Journal of Neuroscience, 37*(40), pp. 9603–9613.

Bonaz, B., Bazin, T. and Pellissier, S., 2018. The vagus nerve at the interface of the microbiota-gut-brain axis. *Frontiers in neuroscience, 12*, p. 49

Bondi, C.O., Semple, B.D., Noble-Haeusslein, L.J., Osier, N.D., Carlson, S.W., Dixon, C.E., Giza, C.C. and Kline, A.E., 2015. Found in translation: Understanding the biology and behavior of experimental traumatic brain injury. *Neuroscience & Biobehavioral Reviews, 58*, pp. 123–146.

Bonnar, D., Bartel, K., Kakoschke, N. and Lang, C., 2018. Sleep interventions designed to improve athletic performance and recovery: a systematic review of current approaches. *Sports medicine, 48*(3), pp. 683–703.

Borovikova, L.V., Ivanova, S., Zhang, M., Yang, H., Botchkina, G.I., Watkins, L.R., Wang, H., Abumrad, N., Eaton, J.W. and Tracey, K.J., 2000. Vagus nerve stimulation attenuates the systemic inflammatory response to endotoxin. *Nature, 405*(6785), pp. 458–462.

Botchway, E.N., Godfrey, C., Ryan, N.P., Hearps, S., Nicholas, C.L., Anderson, V.A. and Catroppa, C., 2020. Sleep disturbances in young adults with childhood traumatic brain injury: relationship with fatigue, depression, and quality of life. *Brain injury, 34*(12), pp. 1579–1589.

Boysen, S.T. 2006. *The impact of symbolic representations on chimpanzee cognition.* In S. Hurley & M. Nudds (Eds.), *Rational animals?* (p. 489–511). Oxford University Press.

Brait, V.H., Arumugam, T.V., Drummond, G.R. and Sobey, C.G., 2012. Importance of T lymphocytes in brain injury, immunodeficiency, and recovery after cerebral ischemia. *Journal of Cerebral Blood Flow & Metabolism, 32*(4), pp. 598–611.

Brandt, J. and Leong, C., 2017. Benzodiazepines and Z-drugs: an updated review of major adverse outcomes reported on in epidemiologic research. *Drugs in R&D, 17*(4), pp. 493–507.

Bravo, J.A., Forsythe, P., Chew, M.V., Escaravage, E., Savignac, H.M., Dinan, T.G., Bienenstock, J. and Cryan, J.F., 2011. Ingestion of Lactobacillus strain regulates emotional behavior and central GABA receptor expression in a mouse via the vagus nerve. *Proceedings of the National Academy of Sciences, 108*(38), pp. 16050–16055

Bremer, F., 1937. Cerebral activity during sleep and narcosis: contribution to the study of the mechanisms of sleep. *Bull Acad Roy Med Bel, 4*, pp. 240–275.

Bretherton, B., Atkinson, L., Murray, A., Clancy, J., Deuchars, S. and Deuchars, J., 2019. Effects of transcutaneous vagus nerve stimulation in individuals aged 55 years or above: potential benefits of daily stimulation. *Aging (Albany NY), 11*(14), p. 4836

Broderick, M., Rosignoli, L., Lunagariya, A. and Nagaraja, N., 2020. Hypertension is a Leading Cause of Nontraumatic Intracerebral Hemorrhage in Young Adults. *Journal of Stroke and Cerebrovascular Diseases, 29*(5), p. 104719.

Brown, E.N., Lydic, R. and Schiff, N.D., 2010. General anesthesia, sleep, and coma. *New England Journal of Medicine, 363*(27), pp. 2638–2650.

Bryan, C.J., 2013. Repetitive traumatic brain injury (or concussion) increases severity of sleep disturbance among deployed military personnel. *Sleep, 36*(6), pp. 941–946.

Burke, T.M., Markwald, R.R., McHill, A.W., Chinoy, E.D., Snider, J.A., Bessman, S.C., Jung, C.M., O'Neill, J.S. and Wright, K.P., 2015. Effects of caffeine on the human circadian clock in vivo and in vitro. *Science translational medicine, 7*(305), pp. 305ra146–305ra146.

Buscemi, N., Vandermeer, B., Pandya, R., Hooton, N., Tjosvold, L., Hartling, L., Baker, G., Vohra, S. and Klassen, T., 2004. Melatonin for treatment of sleep disorders: summary. *AHRQ evidence report summaries.* Available from: https://www.ncbi.nlm.nih.gov/books/NBK11941/

Busek, P. and Faber, J., 2000. The influence of traumatic brain lesion on sleep architecture. *Sbornik lekarsky, 101*(3), pp. 233–239.

Busl, K.M. and Greer, D.M., 2010. Hypoxic-ischemic brain injury: pathophysiology, neuropathology and mechanisms. *NeuroRehabilitation, 26*(1), pp. 5–13.

Büssing, A., Michalsen, A., Khalsa, S.B.S., Telles, S. and Sherman, K.J., 2012. Effects of yoga on mental and physical health: a short summary of reviews. *Evidence-Based Complementary and Alternative Medicine, 2012*, 165410. https://doi.org/10.1155/2012/165410

Butt, M.F., Albusoda, A., Farmer, A.D. and Aziz, Q., 2020. The anatomical basis for transcutaneous auricular vagus nerve stimulation. *Journal of anatomy, 236*(4), pp. 588–611.

Cadosch, D., Toffoli, A.M., Gautschi, O.P., Frey, S.P., Zellweger, R., Skirving, A.P. and Filgueira, L., 2010. Serum after traumatic brain injury increases proliferation and supports expression of osteoblast markers in muscle cells. *JBJS, 92*(3), pp. 645–653.

Calhoun, D.A. and Harding, S.M., 2010. Sleep and hypertension. *Chest, 138*(2), pp. 434–443

Capone, F., Miccinilli, S., Pellegrino, G., Zollo, L., Simonetti, D., Bressi, F., Florio, L., Ranieri, F., Falato, E., Di Santo, A. and Pepe, A., 2017. Transcutaneous vagus nerve stimulation combined with robotic rehabilitation improves upper limb function after stroke. *Neural plasticity, 2017*, 7876507. https://doi.org/10.1155/2017/7876507

Caporro, M., Haneef, Z., Yeh, H.J., Lenartowicz, A., Buttinelli, C., Parvizi, J. and Stern, J.M., 2012. Functional MRI of sleep spindles and K-complexes. *Clinical neurophysiology, 123*(2), pp. 303–309.

Cappuccio, F.P., D'Elia, L., Strazzullo, P. and Miller, M.A., 2010. Quantity and quality of sleep and incidence of type 2 diabetes: a systematic review and meta-analysis. *Diabetes care, 33*(2), pp. 414–420.

Carbonara, M., Fossi, F., Zoerle, T., Ortolano, F., Moro, F., Pischiutta, F., Zanier, E.R. and Stocchetti, N., 2018. Neuroprotection in traumatic brain injury: mesenchymal stromal cells can potentially overcome some limitations of previous clinical trials. *Frontiers in neurology, 9*, p. 885.

Carew, T.J. and Sahley, C.L., 1986. Invertebrate learning and memory: from behavior to molecules. *Annual review of neuroscience, 9*(1), pp. 435–487.

Carneiro-Barrera, A., Díaz-Román, A., Guillén-Riquelme, A. and Buela-Casal, G., 2019. Weight loss and lifestyle interventions for obstructive sleep apnoea in adults: Systematic review and meta-analysis. *Obesity Reviews, 20*(5), pp. 750–762.

Chang, T.I., Lee, U.K., Zeidler, M.R., Liu, S.Y., Polanco, J.C. and Friedlander, A.H., 2019. Severity of obstructive sleep apnea is positively associated with the presence of carotid artery atheromas. *Journal of Oral and Maxillofacial Surgery*, *77*(1), pp. 93–99

Chaudhary, A., Kumari, V. and Neetu, N., 2020. Sleep Promotion among Critically Ill Patients: Earplugs/Eye Mask versus Ocean Sound—A Randomized Controlled Trial Study. *Critical Care Research and Practice*, 2020, 8898172. https://doi.org/1 0.1155/2020/8898172

Chen, L., Deng, C., Chen, X., Zhang, X., Chen, B., Yu, H., Qin, Y., Xiao, K., Zhang, H. and Sun, X., 2020. Ocular manifestations and clinical characteristics of 535 cases of COVID-19 in Wuhan, China: a cross-sectional study. *Acta ophthalmologica*, *98*(8), pp. e951–e959

Cho, C.Y., Kim, N.Y., Kang, J.W., Leem, Y.C., Hong, S.H., Lim, W., Kim, S.T. and Park, S.J., 2013. Improved light extraction efficiency in blue light-emitting diodes by SiO2-coated ZnO nanorod arrays. *Applied Physics Express*, *6*(4), p. 042102

Christensen, J., Wright, D.K., Yamakawa, G.R., Shultz, S.R. and Mychasiuk, R., 2020. Repetitive mild traumatic brain injury alters glymphatic clearance rates in limbic structures of adolescent female rats. *Scientific reports*, *10*(1), pp. 1–9.

Chu, M., Huang, Z.L., Qu, W.M., Eguchi, N., Yao, M.H. and Urade, Y., 2004. Extracellular histamine level in the frontal cortex is positively correlated with the amount of wakefulness in rats. *Neuroscience research*, *49*(4), pp. 417–420.

Clark, J.F., Loftspring, M., Wurster, W.L., Beiler, S., Beiler, C., Wagner, K.R. and Pyne-Geithman, G.J., 2008. Bilirubin oxidation products, oxidative stress, and intracerebral hemorrhage. In *Cerebral Hemorrhage* (pp. 7–12). Vienna: Springer.

Costantini, T.W., Bansal, V., Krzyzaniak, M., Putnam, J.G., Peterson, C.Y., Loomis, W.H., Wolf, P., Baird, A., Eliceiri, B.P. and Coimbra, R., 2010. Vagal nerve stimulation protects against burn-induced intestinal injury through activation of enteric glia cells. *American Journal of Physiology-Gastrointestinal and Liver Physiology*, *299*(6), pp. G1308–G1318.

Costantini, T.W., Krzyzaniak, M., Cheadle, G.A., Putnam, J.G., Hageny, A.M., Lopez, N., Eliceiri, B.P., Bansal, V. and Coimbra, R., 2012. Targeting α-7 nicotinic acetylcholine receptor in the enteric nervous system: a cholinergic agonist prevents gut barrier failure after severe burn injury. *The American journal of pathology*, *181*(2), pp. 478–486

Courville, C.B., 1937. Pathology of the central nervous system, part 4. *Pacific, Mountain View, Calif.*

Cousins, J.N. and Fernández, G., 2019. The impact of sleep deprivation on declarative memory. *Progress in brain research*, *246*, pp. 27–53.

Craik, F.I. and Lockhart, R.S., 1972. Levels of processing: A framework for memory research. *Journal of verbal learning and verbal behavior*, *11*(6), pp. 671–684.

Crompton, M.R., 1971. Brainstem lesions due to closed head injury. *The Lancet*, *297*(7701), pp. 669–673.

Cross, N., Terpening, Z., Rogers, N.L., Duffy, S.L., Hickie, I.B., Lewis, S.J. and Naismith, S.L., 2015. Napping in older people 'at risk' of dementia: relationships with depression, cognition, medical burden and sleep quality. *Journal of sleep research*, *24*(5), pp. 494–502

Crowther, M.E., Ferguson, S.A., Vincent, G.E. and Reynolds, A.C., 2021. Non-Pharmacological Interventions to Improve Chronic Disease Risk Factors and Sleep in Shift Workers: A Systematic Review and Meta-Analysis. *Clocks & sleep*, *3*(1), pp. 132–184.

Culpeper, N., 2006. *Culpeper's complete herbal & English physician.* Applewood Books

Daan, S., Beersma, D.G. and Borbély, A.A., 1984. Timing of human sleep: recovery process gated by a circadian pacemaker. *American Journal of Physiology-Regulatory, Integrative and Comparative Physiology*, *246*(2), pp. R161–R183.

Da Mesquita S., Fu Z., Kipnis J., 2018. The Meningeal Lymphatic System: A New Player in Neurophysiology. *Neuron. 100*(2), pp. 375–388.

Davis, H., Davis, P.A., Loomis, A.L., Harvey, E.N. and Hobart, G., 1937. Changes in human brain potentials during the onset of sleep. *Science, 86*(2237), pp. 448–450.

De Gennaro, L., Ferrara, M., Vecchio, F., Curcio, G. and Bertini, M., 2005. An electroencephalographic fingerprint of human sleep. *Neuroimage, 26*(1), pp. 114–122.

Dejerine, J., 1906. Roussy G. Le syndrome thalamique. *Rev Neural (Paris)*, *1*(2), pp. 521–532.

Del Felice, A., Arcaro, C., Storti, S.F., Fiaschi, A. and Manganotti, P., 2014. Electrical source imaging of sleep spindles. *Clinical EEG and neuroscience, 45*(3), pp. 184–192.

Dement, W. and Kleitman, N., 1957. Cyclic variations in EEG during sleep and their relation to eye movements, body motility, and dreaming. *Electroencephalography and clinical neurophysiology*, *9*(4), pp. 673–690.

Dement, W. and Wolpert, E.A., 1958. The relation of eye movements, body motility, and external stimuli to dream content. *Journal of experimental psychology*, *55*(6), p. 543.

Demoule, A., Carreira, S., Lavault, S., Pallanca, O., Morawiec, E., Mayaux, J., Arnulf, I. and Similowski, T., 2017. Impact of earplugs and eye mask on sleep in critically ill patients: a prospective randomized study. *Critical Care, 21*(1), pp. 1–9.

Diekelmann, S. and Born, J., 2010. The memory function of sleep. *Nature Reviews Neuroscience, 11*(2), pp. 114–126.

Ding, K., Gupta, P.K. and Diaz-Arrastia, R., 2016. Epilepsy after traumatic brain injury. In Laskowitz D. and Grant G. (eds.), *Translational research in traumatic brain injury.* Chapter 14. Boca Raton (FL): CRC Press/Taylor and Francis Group. Available from: https://www.ncbi.nlm.nih.gov/books/NBK326716/

Ding, K., Tarumi, T., Tomoto, T., Bell, K.R., Madden, C., Dieppa, M., Cullum, C.M., Zhang, S. and Zhang, R., 2020. A proof-of-concept trial of a community-based aerobic exercise program for individuals with traumatic brain injury. *Brain injury*, *35*(2), pp. 1–8

Dolcos, F., LaBar, K.S. and Cabeza, R., 2005. Remembering one year later: role of the amygdala and the medial temporal lobe memory system in retrieving emotional memories. *Proceedings of the National Academy of Sciences, 102*(7), pp. 2626–2631.

Doll, D.N., Barr, T.L. and Simpkins, J.W., 2014. Cytokines: their role in stroke and potential use as biomarkers and therapeutic targets. *Aging and disease, 5*(5), p. 294.

Doll, R. and Hill, A.B., 1950. Smoking and carcinoma of the lung. *British medical journal*, *2*(4682), p. 739.

Doll, R. and Hill, A.B., 1954. The mortality of doctors in relation to their smoking habits. *British medical journal*, *1*(4877), p. 1451.

Dominguez-Villar, M. and Hafler, D.A., 2018. Regulatory T cells in autoimmune disease. *Nature immunology*, *19*(7), pp. 665–673.

Du, T., Jing, X., Song, S., Lu, S., Xu, L., Tong, X. and Yan, H., 2020. Therapeutic effect of enteral nutrition supplemented with probiotics in the treatment of severe craniocerebral injury: a systematic review and meta-analysis. *World neurosurgery*, *139*, pp. e553–e571

Dubiel R., Williams B., Sullivan E., et al., 2019 Prevalence of 25-hydroxyvitamin D deficiency in the acute rehabilitation population following traumatic brain injury. *NeuroRehabilitation 45*(4), pp. 513–517

Ducharme-Crevier, L., Press, C.A., Kurz, J.E., Mills, M.G., Goldstein, J.L. and Wainwright, M.S., 2017. Early presence of sleep spindles on electroencephalography is associated with good outcome after pediatric cardiac arrest. *Pediatric critical care medicine*, *18*(5), pp. 452–460.

Duclos, C., Dumont, M., Arbour, C., Paquet, J., Blais, H., Menon, D.K., De Beaumont, L., Bernard, F. and Gosselin, N., 2017. Parallel recovery of consciousness and sleep in acute traumatic brain injury. *Neurology*, *88*(3), pp. 268–275.

Eckert, D.J., 2018. Phenotypic approaches to obstructive sleep apnoea–new pathways for targeted therapy. *Sleep medicine reviews*, *37*, pp. 45–59.

Edwards, B.A., Bristow, C., O'Driscoll, D.M., Wong, A.M., Ghazi, L., Davidson, Z.E., Young, A., Truby, H., Haines, T.P. and Hamilton, G.S., 2019. Assessing the impact of diet, exercise and the combination of the two as a treatment for OSA: A systematic review and meta-analysis. *Respirology*, *24*(8), pp. 740–751.

Eichenbaum, H. and Cohen, N.J., 2004. *From conditioning to conscious recollection: Memory systems of the brain* (No. 35). Oxford University Press on Demand.

Elia, M., Klepper, J., Leiendecker, B. and Hartmann, H., 2017. Ketogenic diets in the treatment of epilepsy. *Current pharmaceutical design*, *23*(37), pp. 5691–5701

Elliott, A.S., Huber, J.D., O'Callaghan, J.P., Rosen, C.L. and Miller, D.B., 2014. A review of sleep deprivation studies evaluating the brain transcriptome. *SpringerPlus*, *3*(1), pp. 1–12.

Elliott J.E., Opel R.A., Weymann K.B., Chau A.Q., Papesh M.A., Callahan M.L., Storzbach D., Lim M.M., 2018 Jul 15. Sleep Disturbances in Traumatic Brain Injury: Associations With Sensory Sensitivity. *Journal of Clinical Sleep Medicine*, *14*(7), pp. 1177–1186.

El-Khatib, H., Arbour, C., Sanchez, E., Dumont, M., Duclos, C., Blais, H., Carrier, J., Paquet, J. and Gosselin, N., 2019. Towards a better understanding of increased sleep duration in the chronic phase of moderate to severe traumatic brain injury: an actigraphy study. *Sleep medicine*, *59*, pp. 67–75.

Elvers, K.T., Wilson, V.J., Hammond, A., Duncan, L., Huntley, A.L., Hay, A.D. and van der Werf, E.T., 2020. Antibiotic-induced changes in the human gut microbiota for the most commonly prescribed antibiotics in primary care in the UK: a systematic review. *BMJ open*, *10*(9), p. e035677.

Endo, T., Matsumura, R., Tokuda, I.T., Yoshikawa, T., Shigeyoshi, Y., Node, K., Sakoda, S. and Akashi, M., 2020. Bright light improves sleep in patients with parkinson's disease: possible role of circadian restoration. *Scientific reports*, *10*(1), pp. 1–10

Eskelinen, M.H. and Kivipelto, M., 2010. Caffeine as a protective factor in dementia and Alzheimer's disease. *Journal of Alzheimer's Disease*, *20*(s1), pp. S167–S174.

Ezz, H.S.A., Noor, A.E., Mourad, I.M., Fahmy, H. and Khadrawy, Y.A., 2021. Neurochemical effects of sleep deprivation in the hippocampus of the pilocarpine-induced rat model of epilepsy. *Iranian Journal of Basic Medical Sciences*, *24*(1), p. 85.

Fang, Z., Ray, L.B., Owen, A.M. and Fogel, S.M., 2019. Brain activation time-locked to sleep spindles associated with human cognitive abilities. *Frontiers in neuroscience*, *13*, p. 46.

Faraut, B., Boudjeltia, K.Z., Dyzma, M., Rousseau, A., David, E., Stenuit, P., Franck, T., Van Antwerpen, P., Vanhaeverbeek, M. and Kerkhofs, M., 2011. Benefits of napping and an extended duration of recovery sleep on alertness and immune cells after acute sleep restriction. *Brain, behavior, and immunity*, *25*(1), pp. 16–24.

Fedele, B., Williams, G., McKenzie, D., Sutherland, E. and Olver, J., 2020. Subacute sleep disturbance in moderate to severe traumatic brain injury: a systematic review. *Brain injury*, *34*(3), pp. 316–327.

Fein M., 2020. A new dementia treatment with quieting focus, subtle sound vibration, and intentional shared silence: Introducing Resonant Silence Technique: Innovative practice. *Dementia*, *19*(3), pp. 894–898.

Feng, W., Bowden, M.G. and Kautz, S., 2013. Review of transcranial direct current stimulation in poststroke recovery. *Topics in stroke rehabilitation*, *20*(1), pp. 68–77.

Fernandez, L.M. and Lüthi, A., 2020. Sleep spindles: mechanisms and functions. *Physiological reviews*, *100*(2), pp. 805–868

Ferracioli-Oda, E., Qawasmi, A. and Bloch, M.H., 2013. Meta-analysis: melatonin for the treatment of primary sleep disorders. *PloS one*, *8*(5), p. e63773.

Ferre, A., Ribó, M., Rodriguez-Luna, D., Romero, O., Sampol, G., Molina, C.A. and Álvarez-Sabin, J., 2013. Strokes and their relationship with sleep and sleep disorders. *Neurologia (English Edition)*, *28*(2), pp. 103–118.

Fismer, K.L. and Pilkington, K., 2012. Lavender and sleep: A systematic review of the evidence. *European Journal of Integrative Medicine*, *4*(4), pp. e436–e447

Ford, M.E., Groet, E., Daams, J.G., Geurtsen, G.J., Van Bennekom, C.A. and Van Someren, E.J., 2020. Non-pharmacological treatment for insomnia following acquired brain injury: A systematic review. *Sleep medicine reviews*, *50*, p. 101255

Foster, R.G., Provencio, I., Hudson, D., Fiske, S., De Grip, W. and Menaker, M., 1991. Circadian photoreception in the retinally degenerate mouse (rd/rd). *Journal of Comparative Physiology A*, *169*(1), pp. 39–50.

France, K.G., McLay, L.K., Hunter, J.E. and France, M.L., 2018. Empirical research evaluating the effects of non-traditional approaches to enhancing sleep in typical and clinical children and young people. *Sleep medicine reviews*, *39*, pp. 69–81.

Frieboes, R.M., Müller, U., Murck, H., von Cramon, D.Y., Holsboer, F. and Steiger, A., 1999. Nocturnal hormone secretion and the sleep EEG in patients several months after traumatic brain injury. *The Journal of neuropsychiatry and clinical neurosciences*, *11*(3), pp. 354–360.

Fuller, P., Sherman, D., Pedersen, N.P., Saper, C.B. and Lu, J., 2011. Reassessment of the structural basis of the ascending arousal system. *Journal of Comparative Neurology*, *519*(5), pp. 933–956.

Furlong, E.E., Boose, K.J. and Boysen, S.T., 2008. Raking it in: the impact of enculturation on chimpanzee tool use. *Animal cognition*, *11*(1), pp. 83–97.

Gaberel, T., Gakuba, C., Goulay, R., De Lizarrondo, S.M., Hanouz, J.L., Emery, E., Touze, E., Vivien, D. and Gauberti, M., 2014. Impaired glymphatic perfusion after strokes revealed by contrast-enhanced MRI: a new target for fibrinolysis? *Stroke*, *45*(10), pp. 3092–3096.

Gais, S., Albouy, G., Boly, M., Dang-Vu, T.T., Darsaud, A., Desseilles, M., Rauchs, G., Schabus, M., Sterpenich, V., Vandewalle, G. and Maquet, P., 2007. Sleep transforms the cerebral trace of declarative memories. *Proceedings of the National Academy of Sciences*, *104*(47), pp. 18778–18783.

Gao, Q., Kou, T., Zhuang, B., Ren, Y., Dong, X. and Wang, Q., 2018. The association between vitamin D deficiency and sleep disorders: a systematic review and meta-analysis. *Nutrients*, *10*(10), p. 1395.

Garcia-Cazarin, M.L., Wambogo, E.A., Regan, K.S. and Davis, C.D., 2014. Dietary supplement research portfolio at the NIH, 2009–2011. *The Journal of nutrition*, *144*(4), pp. 414–418.

Gardani, M., Morfiri, E., Thomson, A., O'Neill, B. and McMillan, T.M., 2015. Evaluation of sleep disorders in patients with severe traumatic brain injury during rehabilitation. *Archives of physical medicine and rehabilitation*, *96*(9), pp. 1691–1697.

Gennarelli, T.A., Thibault, L.E., Adams, J.H., Graham, D.I., Thompson, C.J. and Marcincin, R.P., 1982. Diffuse axonal injury and traumatic coma in the primate. *Annals of Neurology: Official Journal of the American Neurological Association and the Child Neurology Society*, *12*(6), pp. 564–574.

Geoffroy, P.A., Schroder, C.M., Reynaud, E. and Bourgin, P., 2019. Efficacy of light therapy versus antidepressant drugs, and of the combination versus monotherapy, in major depressive episodes: A systematic review and meta-analysis. *Sleep medicine reviews*, *48*, p. 101213

George, C.F., 2004. Sleep· 5: Driving and automobile crashes in patients with obstructive sleep apnoea/hypopnoea syndrome. *Thorax*, *59*(9), pp. 804–807.

Gerbarg, P.L., Gootjes, L. and Brown, R.P., 2014. Mind-Body Practices and the Neuro-psychology of Wellbeing. In *Religion and Spirituality Across Cultures* (pp. 227–246). Dordrecht: Springer.

Germain, A., 2013. Sleep disturbances as the hallmark of PTSD: where are we now? *American Journal of Psychiatry*, *170*(4), pp. 372–382.

Girolamo, F., Coppola, C. and Ribatti, D., 2017. Immunoregulatory effect of mast cells influenced by microbes in neurodegenerative diseases. *Brain, behavior, and immunity*, *65*, pp. 68–89.

Giuditta, A., 2014. Sleep memory processing: the sequential hypothesis. *Frontiers in systems neuroscience*, *8*, p. 219.

Gjerstad, M.D., Alves, G. and Maple-Grødem, J., 2018. Excessive daytime sleepiness and REM sleep behavior disorders in parkinson's disease: a narrative review on early intervention with implications to neuroprotection. *Frontiers in neurology*, *9*, p. 961.

Gluck, M., Mercado, E., Myers, C., 2020, *Learning and Memory: from brain to behaviour*, 4th edn, Worth Publications.

Gkotzamanis, V. and Panagiotakos, D., 2020. Dietary interventions and cognition: A systematic review of clinical trials. *Psychiatriki*. *31*, pp. 248–256

Gómez-Pinilla, F., 2008. Brain foods: the effects of nutrients on brain function. *Nature reviews neuroscience, 9*(7), pp. 568–578.

Gothe, N.P. and McAuley, E., 2015. Yoga and cognition: a meta-analysis of chronic and acute effects. *Psychosomatic medicine, 77*(7), pp. 784–797

Gottlieb, D.J., Ellenbogen, J.M., Bianchi, M.T. and Czeisler, C.A., 2018. Sleep deficiency and motor vehicle crash risk in the general population: a prospective cohort study. *BMC medicine, 16*(1), pp. 1–10.

Gray, S.L., Anderson, M.L., Dublin, S., Hanlon, J.T., Hubbard, R., Walker, R., Yu, O., Crane, P.K. and Larson, E.B., 2015. Cumulative use of strong anticholinergics and incident dementia: a prospective cohort study. *JAMA internal medicine, 175*(3), pp. 401–407.

Gray, E.L., McKenzie, D.K. and Eckert, D.J., 2017. Obstructive sleep apnea without obesity is common and difficult to treat: evidence for a distinct pathophysiological phenotype. *Journal of Clinical Sleep Medicine, 13*(1), pp. 81–88.

Griefahn, B. and Basner, M., 2011. Disturbances of sleep by noise. *Proceedings of the Acoustics, Gold Coast, Australia*, pp. 2–4.

Grima, N.A., Ponsford, J.L. and Pase, M.P., 2017. Sleep complications following traumatic brain injury. *Current opinion in pulmonary medicine, 23*(6), pp. 493–499.

Grima, N.A., Rajaratnam, S.M., Mansfield, D., Sletten, T.L., Spitz, G. and Ponsford, J.L., 2018. Efficacy of melatonin for sleep disturbance following traumatic brain injury: a randomised controlled trial. *BMC medicine, 16*(1), pp. 1–10.

Gründer, G. and Cumming, P., 2016. The dopamine hypothesis of schizophrenia: Current status. In *The neurobiology of schizophrenia* (pp. 109–124). Academic Press.

Guadagno, J.V., Calautti, C. and Baron, J.C., 2003. Progress in imaging stroke: emerging clinical applications. *British medical bulletin, 65*(1), pp. 145–157.

Guerriero, R.M., Giza, C.C. and Rotenberg, A., 2015. Glutamate and GABA imbalance following traumatic brain injury. *Current neurology and neuroscience reports, 15*(5), p. 27.

Gunduz, S., Kosger, H., Aldemir, S., Akcal, B., Tevrizci, H., Hizli, D. and Celik, H.T., 2016. Sleep deprivation in the last trimester of pregnancy and inadequate vitamin D: Is there a relationship? *Journal of the Chinese Medical Association, 79*(1), pp. 34–38

Gumenyuk, V., Roth, T., Moran, J.E., Jefferson, C., Bowyer, S.M., Tepley, N. and Drake, C.L., 2009. Cortical locations of maximal spindle activity: magnetoencephalography (MEG) study. *Journal of sleep research, 18*(2), pp. 245–253.

Gustafsson, N., Ahlqvist, J.B., Näslund, U., Wester, P., Buhlin, K., Gustafsson, A. and Jäghagen, E.L., 2018. Calcified carotid artery atheromas in panoramic radiographs are associated with a first myocardial infarction: a case-control study. *Oral surgery, oral medicine, oral pathology and oral radiology, 125*(2), pp. 199–204.

Hakon, J., Moghiseh, M., Poulsen, I., Øland, C.M., Hansen, C.P. and Sabers, A., 2020. Transcutaneous vagus nerve stimulation in patients with severe traumatic brain injury: a feasibility trial. *Neuromodulation: Technology at the Neural Interface, 23*(6), pp. 859–864

Hammond, F.M., Corrigan, J.D., Ketchum, J.M., Malec, J.F., Dams-O'Conner, K., Hart, T., Novack, T.A., Bogner, J., Dahdah, M.N. and Whiteneck, G.G., 2019. Prevalence of medical and psychiatric comorbidities following traumatic brain injury. *The Journal of head trauma rehabilitation, 34*(4), p. E1.

Harriott, A.M., Karakaya, F. and Ayata, C., 2020. Headache after ischemic stroke: A systematic review and meta-analysis. *Neurology*, *94*(1), pp. e75–e86.

Harrison, R.A. and Field, T.S., 2015. Post stroke pain: identification, assessment, and therapy. *Cerebrovascular diseases*, *39*(3–4), pp. 190–201.

Harrison, Y. and Horne, J.A., 2000. Sleep loss and temporal memory. *The Quarterly Journal of Experimental Psychology: Section A*, *53*(1), pp. 271–279.

Harrison-Felix, C., Whiteneck, G., DeVivo, M.J., Hammond, F.M. and Jha, A., 2006. Causes of death following 1 year postinjury among individuals with traumatic brain injury. *The Journal of head trauma rehabilitation*, *21*(1), pp. 22–33.

Hayashi, M., Motoyoshi, N. and Hori, T., 2005. Recuperative power of a short daytime nap with or without stage 2 sleep. *Sleep*, *28*(7), pp. 829–836

Headway U.K., 2019 Survey of Fatigue. Headway Publications.

Henninger, N. and Fisher, M., 2016. Extending the time window for endovascular and pharmacological reperfusion. *Translational Stroke Research*, *7*(4), pp. 284–293.

Henry, R.J., Ritzel, R.M., Barrett, J.P., Doran, S.J., Jiao, Y., Leach, J.B., Szeto, G.L., Wu, J., Stoica, B.A., Faden, A.I. and Loane, D.J., 2020. Microglial depletion with CSF1R inhibitor during chronic phase of experimental traumatic brain injury reduces neurodegeneration and neurological deficits. *Journal of Neuroscience*, *40*(14), pp. 2960–2974.

Hermann, D.M. and Bassetti, C.L., 2009. Sleep-related breathing and sleep-wake disturbances in ischemic stroke. *Neurology*, *73*(16), pp. 1313–1322.

Herron, K., Dijk, D.J., Dean, P., Seiss, E. and Sterr, A., 2014. Quantitative electroencephalography and behavioural correlates of daytime sleepiness in chronic stroke. *BioMed research international*, *2014*, 794086. https://doi.org/10.1155/2014/794086

Hintze, J.P. and Edinger, J.D., 2018. Hypnotic discontinuation in chronic insomnia. *Sleep medicine clinics*, *13*(2), pp. 263–270.

Hirshkowitz, M., Whiton, K., Albert, S.M., Alessi, C., Bruni, O., DonCarlos, L., Hazen, N., Herman, J., Hillard, P.J.A., Katz, E.S. and Kheirandish-Gozal, L., 2015. National Sleep Foundation's updated sleep duration recommendations. *Sleep health*, *1*(4), pp. 233–243.

Hoffmann, M. and Schmitt, F., 2004. Cognitive impairment in isolated subtentorial stroke. *Acta Neurologica Scandinavica*, *109*(1), pp. 14–24.

Hoge, C.W., McGurk, D., Thomas, J.L., Cox, A.L., Engel, C.C. and Castro, C.A., 2008. Mild traumatic brain injury in US soldiers returning from Iraq. *New England journal of medicine*, *358*(5), pp. 453–463.

Holth, J.K., Fritschi, S.K., Wang, C., Pedersen, N.P., Cirrito, J.R., Mahan, T.E., Finn, M.B., Manis, M., Geerling, J.C., Fuller, P.M. and Lucey, B.P., 2019. The sleep-wake cycle regulates brain interstitial fluid tau in mice and CSF tau in humans. *Science*, *363*(6429), pp. 880–884.

Hou, L., Han, X., Sheng, P., Tong, W., Li, Z., Xu, D., Yu, M., Huang, L., Zhao, Z., Lu, Y. and Dong, Y., 2013. Risk factors associated with sleep disturbance following traumatic brain injury: clinical findings and questionnaire-based study. *PLoS One*, *8*(10), p. e76087.

Howden, E.J., Sarma, S., Lawley, J.S., Opondo, M., Cornwell, W., Stoller, D., Urey, M.A., Adams-Huet, B. and Levine, B.D., 2018. Reversing the cardiac effects of sedentary aging in middle age—a randomized controlled trial: implications for heart failure prevention. *Circulation*, *137*(15), pp. 1549–1560.

Howes, O., McCutcheon, R. and Stone, J., 2015. Glutamate and dopamine in schizophrenia: an update for the 21st century. *Journal of psychopharmacology*, *29*(2), pp. 97–115.

Huang, H., Cheng, W.X., Hu, Y.P., Chen, J.H., Zheng, Z.T. and Zhang, P., 2018. Relationship between heterotopic ossification and traumatic brain injury: why severe traumatic brain injury increases the risk of heterotopic ossification. *Journal of orthopaedic translation*, *12*, pp. 16–25.

Huang, R., Wang, K. and Hu, J., 2016. Effect of probiotics on depression: a systematic review and meta-analysis of randomized controlled trials. *Nutrients*, *8*(8), p. 483

Huang, F.P., Xi, G., Keep, R.F., Hua, Y., Nemoianu, A. and Hoff, J.T., 2002. Brain edema after experimental intracerebral hemorrhage: role of hemoglobin degradation products. *Journal of neurosurgery*, *96*(2), pp. 287–293.

Hwang, S.T., Stevens, S.J., Fu, A.X. and Proteasa, S.V., 2019. Intractable generalized epilepsy: therapeutic approaches. *Current neurology and neuroscience reports*, *19*(4), pp. 1–10.

Ikram, M.A., Wieberdink, R.G. and Koudstaal, P.J., 2012. International epidemiology of intracerebral hemorrhage. *Current atherosclerosis reports*, *14*(4), pp. 300–306.

Imbach, L.L., Büchele, F., Valko, P.O., Li, T., Maric, A., Stover, J.F., Bassetti, C.L., Mica, L., Werth, E. and Baumann, C.R., 2016. Sleep–wake disorders persist 18 months after traumatic brain injury but remain underrecognized. *Neurology*, *86*(21), pp. 1945–1949.

Irwin, M.R., Cole, J.C. and Nicassio, P.M., 2006. Comparative meta-analysis of behavioral interventions for insomnia and their efficacy in middle-aged adults and in older adults 55+ years of age. *Health Psychology*, *25*(1), p. 3.

Ioannides, A.A., Kostopoulos, G.K., Liu, L. and Fenwick, P.B., 2009. MEG identifies dorsal medial brain activations during sleep. *Neuroimage*, *44*(2), pp. 455–468.

Issek, V. and Suchan, B., 2021. Preventing dementia? Interventional approaches in mild cognitive impairment. Neuroscience & Biobehavioural Reviews. **Vol 122**, pp. 143–164

Ito, M., Komai, K., Mise-Omata, S., Iizuka-Koga, M., Noguchi, Y., Kondo, T., Sakai, R., Matsuo, K., Nakayama, T., Yoshie, O. and Nakatsukasa, H., 2019. Brain regulatory T cells suppress astrogliosis and potentiate neurological recovery. *Nature*, *565*(7738), pp. 246–250.

Jacobson, L.H., Callander, G.E. and Hoyer, D., 2014. Suvorexant for the treatment of insomnia. *Expert review of clinical pharmacology*, *7*(6), pp. 711–773

Jaeger, B.N., Linker, S.B., Parylak, S.L., Barron, J.J., Gallina, I.S., Saavedra, C.D., Fitzpatrick, C., Lim, C.K., Schafer, S.T., Lacar, B. and Jessberger, S., 2018. A novel environment-evoked transcriptional signature predicts reactivity in single dentate granule neurons. *Nature communications*, *9*(1), pp. 1–15.

Jamall, O.A., Feeney, C., Zaw-Linn, J., Malik, A., Niemi, M.E., Tenorio-Jimenez, C., Ham, T.E., Jilka, S.R., Jenkins, P.O., Scott, G. and Li, L.M., 2016. Prevalence and correlates of vitamin D deficiency in adults after traumatic brain injury. *Clinical endocrinology*, *85*(4), pp. 636–644.

Jassam, Y.N., Izzy, S., Whalen, M., McGavern, D.B. and El Khoury, J., 2017. Neuroimmunology of traumatic brain injury: time for a paradigm shift. *Neuron*, *95*(6), pp. 1246–1265.

Jaster, J.H., 2018. Reperfusion injury to ischemic medullary brain nuclei after stopping continuous positive airway pressure-induced CO2-reduced vasoconstriction in sleep apnea. *Journal of thoracic disease*, *10*(Suppl 17), p. S2029.

Javidi, E. and Magnus, T., 2019. Autoimmunity after ischemic stroke and brain injury. *Frontiers in immunology*, *10*, p. 686.

Jehan, S., Farag, M., Zizi, F., Pandi-Perumal, S.R., Chung, A. and Truong, A., 2018. Obstructive sleep apnea and stroke. *Sleep medicine and disorders: international journal*, *2*(5), p. 120.

Jenkins, J.G. and Dallenbach, K.M., 1924. Obliviscence during sleep and waking. *The American Journal of Psychology*, *35*(4), pp. 605–612.

Jennett, B. and Miller, J.D., 1972. Infection after depressed fracture of skull: Implications for management of nonmissile injuries. *Journal of neurosurgery*, *36*(3), pp. 333–339.

Jiménez-Anguiano, A., García-García, F., Mendoza-Ramírez, J., Durán-Vázquez, A. and Drucker-Colín, R., 1996. Brain distribution of vasoactive intestinal peptide receptors following REM sleep deprivation. *Brain research*, *728*(1), pp. 37–46.

Jiruska, P., De Curtis, M., Jefferys, J.G., Schevon, C.A., Schiff, S.J. and Schindler, K., 2013. Synchronization and desynchronization in epilepsy: controversies and hypotheses. *The Journal of physiology*, *591*(4), pp. 787–797

Johansson, B., Andréll, P., Rönnbäck, L. and Mannheimer, C., 2020. Follow-up after 5.5 years of treatment with methylphenidate for mental fatigue and cognitive function after a mild traumatic brain injury. *Brain injury*, *34*(2), pp. 229–235.

Johns, M.W., 1991. A new method for measuring daytime sleepiness: the Epworth sleepiness scale. *sleep*, *14*(6), pp. 540–545.

Johnson, V.E., Stewart, J.E., Begbie, F.D., Trojanowski, J.Q., Smith, D.H. and Stewart, W., 2013. Inflammation and white matter degeneration persist for years after a single traumatic brain injury. *Brain*, *136*(1), pp. 28–42.

Jouvet, M., 1999. *The paradox of sleep: The story of dreaming*. MIT press.

Juengst, S.B., Kumar, R.G., Arenth, P.M. and Wagner, A.K., 2014. Exploratory associations with tumor necrosis factor-α, disinhibition and suicidal endorsement after traumatic brain injury. *Brain, behavior, and immunity*, *41*, pp. 134–143.

Jung, C., Hugot, J.-H. and Barreau., 2010. F. Peyer's Patches: The Immune Sensors of the Intestine. *International Journal of Inflammation*, Article ID 823710.

Kaczmarczyk, R., Tejera, D., Simon, B.J. and Heneka, M.T., 2018. Microglia modulation through external vagus nerve stimulation in a murine model of Alzheimer's disease. *Journal of neurochemistry*, *146*(1), pp. 76–85

Kalyani, B.G., Venkatasubramanian, G., Arasappa, R., Rao, N.P., Kalmady, S.V., Behere, R.V., Rao, H., Vasudev, M.K. and Gangadhar, B.N., 2011. Neurohemodynamic correlates of 'OM'chanting: a pilot functional magnetic resonance imaging study. *International journal of yoga*, *4*(1), pp. 1–3.

Kamalakannan, S.K., Gudlavalleti, A.S., Gudlavalleti, V.S.M., Goenka, S. and Kuper, H., 2015. Challenges in understanding the epidemiology of acquired brain injury in India. *Annals of Indian Academy of Neurology*, *18*(1), p. 66.

Kaniusas, E., Kampusch, S., Tittgemeyer, M., Panetsos, F., Gines, R.F., Papa, M., Kiss, A., Podesser, B., Cassara, A.M., Tanghe, E. and Samoudi, A.M., 2019. Current directions in the auricular vagus nerve stimulation I–a physiological perspective. *Frontiers in neuroscience*, *13*, p. 854

Kaniusas, E., Kampusch, S., Tittgemeyer, M., Panetsos, F., Gines, R.F., Papa, M., Kiss, A., Podesser, B., Cassara, A.M., Tanghe, E. and Samoudi, A.M., 2019. Current directions in the auricular vagus nerve stimulation II–an engineering perspective. *Frontiers in neuroscience, 13*, p. 772.

Kanti Das, T., Wati, M.R. and Fatima-Shad, K., 2015. Oxidative stress gated by Fenton and Haber Weiss reactions and its association with Alzheimer's disease. *Archives of Neuroscience, 2*(2), e60038. doi: 10.5812/archneurosci.20078.

Karaszewski, B., Carpenter, T.K., Thomas, R.G., Armitage, P.A., Lymer, G.K.S., Marshall, I., Dennis, M.S. and Wardlaw, J.M., 2013. Relationships between brain and body temperature, clinical and imaging outcomes after ischemic stroke. *Journal of Cerebral Blood Flow & Metabolism, 33*(7), pp. 1083–1089.

Katz, D.A. and McHorney, C.A., 1998. Clinical correlates of insomnia in patients with chronic illness. *Archives of internal medicine, 158*(10), pp. 1099–1107.

Kempf, J., Werth, E., Kaiser, P.R., Bassetti, C.L. and Baumann, C.R., 2010. Sleep–wake disturbances 3 years after traumatic brain injury. *Journal of Neurology, Neurosurgery & Psychiatry, 81*(12), pp. 1402–1405.

Kerr, N.A., de Rivero Vaccari, J.P., Weaver, C., Dietrich, W.D., Ahmed, T. and Keane, R.W., 2021. Enoxaparin attenuates acute lung injury and inflammasome activation after traumatic brain injury. *Journal of neurotrauma, 38*(5), pp. 646–654

Killgore, W.D., Vanuk, J.R., Shane, B.R., Weber, M. and Bajaj, S., 2020. A randomized, double-blind, placebo-controlled trial of blue wavelength light exposure on sleep and recovery of brain structure, function, and cognition following mild traumatic brain injury. *Neurobiology of disease, 134*, p. 104679.

Kim, H., Im, S., Kim, Y., Sohn, M.K. and Jee, S., 2019. Improvement of cognitive function after continuous positive airway pressure treatment for subacute stroke patients with obstructive sleep apnea: a randomized controlled trial. *Brain sciences, 9*(10), p. 252.

Kim, J.S., Kim, O.L., Seo, W.S., Koo, B.H., Joo, Y. and Bai, D.S., 2009. Memory dysfunctions after mild and moderate traumatic brain injury: comparison between patients with and without frontal lobe injury. *Journal of Korean Neurosurgical Society, 46*(5), p. 459.

Kim, H.A., Perrelli, A., Ragni, A., Retta, F., De Silva, T.M., Sobey, C.G. and Retta, S.F., 2020. Vitamin D deficiency and the risk of cerebrovascular disease. *Antioxidants, 9*(4), p. 327.

Kirsch, D.L., 2010. Low level brain stimulation for anxiety: a review of 50 years of research and supporting data.

Kleitman, N., 1987. *Sleep and wakefulness*. University of Chicago Press.

Klerman, E.B., Wang, W., Duffy, J.F., Dijk, D.J., Czeisler, C.A. and Kronauer, R.E., 2013. Survival analysis indicates that age-related decline in sleep continuity occurs exclusively during NREM sleep. *Neurobiology of aging, 34*(1), pp. 309–318

Klit, H., Finnerup, N.B., Overvad, K., Andersen, G. and Jensen, T.S., 2011. Pain following stroke: a population-based follow-up study. *PloS one, 6*(11), p. e27607.

Knapp, P., Dunn-Roberts, A., Sahib, N., Cook, L., Astin, F., Kontou, E. and Thomas, S.A., 2020. Frequency of anxiety after stroke: An updated systematic review and meta-analysis of observational studies. *International Journal of Stroke, 15*(3), pp. 244–255.

Koenigs, M., Holliday, J., Solomon, J. and Grafman, J., 2010. Left dorsomedial frontal brain damage is associated with insomnia. *Journal of Neuroscience*, *30*(47), pp. 16041–16043.

Kok, B.E., Coffey, K.A., Cohn, M.A., Catalino, L.I., Vacharkulksemsuk, T., Algoe, S.B., Brantley, M. and Fredrickson, B.L., 2013. How positive emotions build physical health: Perceived positive social connections account for the upward spiral between positive emotions and vagal tone. *Psychological science*, *24*(7), pp. 1123–1132

Kondziella, D., 2017. The top 5 neurotransmitters from a clinical neurologist's perspective. *Neurochemical research*, *42*(6), pp. 1767–1771.

Korkmaz, O.T., Tunçel, N., Tunçel, M., Öncü, E.M., Şahintürk, V. and Çelik, M., 2010. Vasoactive intestinal peptide (VIP) treatment of Parkinsonian rats increases thalamic gamma-aminobutyric acid (GABA) levels and alters the release of nerve growth factor (NGF) by mast cells. *Journal of molecular neuroscience*, *41*(2), pp. 278–287.

Kovacic, K., Hainsworth, K., Sood, M., Chelimsky, G., Unteutsch, R., Nugent, M., Simpson, P. and Miranda, A., 2017. Neurostimulation for abdominal pain-related functional gastrointestinal disorders in adolescents: a randomised, double-blind, sham-controlled trial. *The Lancet Gastroenterology & Hepatology*, *2*(10), pp. 727–737

Krapf, H., Widder, B. and Skalej, M., 1998. Small rosarylike infarctions in the centrum ovale suggest hemodynamic failure. *American journal of neuroradiology*, *19*(8), pp. 1479–1484.

Kräuchi, K., Fattori, E., Giordano, A., Falbo, M., Iadarola, A., Aglì, F., Tribolo, A., Mutani, R. and Cicolin, A., 2018. Sleep on a high heat capacity mattress increases conductive body heat loss and slow wave sleep. *Physiology & behavior*, *185*, pp. 23–30

Kreutzmann, J.C., Havekes, R., Abel, T. and Meerlo, P., 2015. Sleep deprivation and hippocampal vulnerability: changes in neuronal plasticity, neurogenesis and cognitive function. *Neuroscience*, *309*, pp. 173–190.

Kumar, R.G., Diamond, M.L., Boles, J.A., Berger, R.P., Tisherman, S.A., Kochanek, P.M. and Wagner, A.K., 2015. Acute CSF interleukin-6 trajectories after TBI: associations with neuroinflammation, polytrauma, and outcome. *Brain, behavior, and immunity*, *45*, pp. 253–262.

Kuybu, O., Tadi, P. and Dossani, R.H., 2020. Posterior cerebral artery stroke. [Updated 2021 Jan 31]. In *StatPearls [Internet]*. Treasure Island (FL): StatPearls Publishing. Available from: https://www.ncbi.nlm.nih.gov/books/NBK532296/

Lacerte, M. and Mesfin, F.B., 2020. Hypoxic Brain Injury. In *StatPearls [Online]*. StatPearls Publishing, Treasure Island (FL). PMID: 30725995.

Lappin, J.M., Darke, S. and Farrell, M., 2017. Stroke and methamphetamine use in young adults: a review. *Journal of Neurology, Neurosurgery & Psychiatry*, *88*(12), pp. 1079–1091.

Lari, Z.N., Hajimonfarednejad, M., Riasatian, M., Abolhassanzadeh, Z., Iraji, A., Vojoud, M., Heydari, M. and Shams, M., 2020. Efficacy of inhaled Lavandula angustifolia Mill. Essential oil on sleep quality, quality of life and metabolic control in patients with diabetes mellitus type II and insomnia. *Journal of ethnopharmacology*, *251*, p. 112560

Laufs, H., Walker, M.C. and Lund, T.E., 2007. 'Brain activation and hypothalamic functional connectivity during human non-rapid eye movement sleep: an EEG/fMRI study'—its limitations and an alternative approach. *Brain*, *130*(7), p. e75

Lee, J.M., Park, J.M., Song, M.K., Oh, Y.J., Kim, C.J. and Kim, Y.J., 2017. The ameliorative effects of exercise on cognitive impairment and white matter injury from blood-brain barrier disruption induced by chronic cerebral hypoperfusion in adolescent rats. *Neuroscience letters*, *638*, pp. 83–89.

Lee, K.A. and Gay, C.L., 2011. Can modifications to the bedroom environment improve the sleep of new parents? Two randomized controlled trials. *Research in nursing & health*, *34*(1), pp. 7–19

Lee, S.Y., Park, S.H., Chung, C., Kim, J.J., Choi, S.Y. and Han, J.S., 2015. Oxytocin protects hippocampal memory and plasticity from uncontrollable stress. *Scientific reports*, *5*(1), pp. 1–9.

Lenzlinger, P.M., Saatman, K.E., Hoover, R.C., Cheney, J.A., Bareyre, F.M., Raghupathi, R., Arnold, L.D. and McIntosh, T.K., 2004. Inhibition of vascular endothelial growth factor receptor (VEGFR) signaling by BSF476921 attenuates regional cerebral edema following traumatic brain injury in rats. *Restorative neurology and neuroscience*, *22*(2), pp. 73–79.

Lerman, I., Davis, B., Huang, M., Huang, C., Sorkin, L., Proudfoot, J., Zhong, E., Kimball, D., Rao, R., Simon, B. and Spadoni, A., 2019. Noninvasive vagus nerve stimulation alters neural response and physiological autonomic tone to noxious thermal challenge. *PloS one*, *14*(2), p. e0201212.

Levenson, J.M. and Sweatt, J.D., 2005. Epigenetic mechanisms in memory formation. *Nature Reviews Neuroscience*, *6*(2), pp. 108–118.

Li, S., Zhai, X., Rong, P., McCabe, M.F., Wang, X., Zhao, J., Ben, H. and Wang, S., 2014. Therapeutic effect of vagus nerve stimulation on depressive-like behavior, hyperglycemia and insulin receptor expression in Zucker fatty rats. *PLoS One*, *9*(11), p. e112066.

Licitra, G., Fredianelli, L., Petri, D. and Vigotti, M.A., 2016. Annoyance evaluation due to overall railway noise and vibration in Pisa urban areas. *Science of the total environment*, *568*, pp. 1315–1325.

Lieberman, D.A., 2020. *Learning and Memory*. 2nd edn. Cambridge: Cambridge University Press. doi: 10.1017/9781108553179.

Lieverse, R., Van Someren, E.J., Nielen, M.M., Uitdehaag, B.M., Smit, J.H. and Hoogendijk, W.J., 2011. Bright light treatment in elderly patients with non-seasonal major depressive disorder: a randomized placebo-controlled trial. *Archives of general psychiatry*, *68*(1), pp. 61–70.

Lim, J. and Dinges, D.F., 2010. A meta-analysis of the impact of short-term sleep deprivation on cognitive variables. *Psychological bulletin*, *136*(3), p. 375.

Lin, J.B., Gerratt, B.W., Bassi, C.J. and Apte, R.S., 2017. Short-wavelength light-blocking eyeglasses attenuate symptoms of eye fatigue. *Investigative ophthalmology & visual science*, *58*(1), pp. 442–447.

Lin, J.S., Sakai, K., Vanni-Mercier, G. and Jouvet, M., 1989. A critical role of the posterior hypothalamus in the mechanisms of wakefulness determined by micro-injection of muscimol in freely moving cats. *Brain research*, *479*(2), pp. 225–240.

Lin, C., Zhao, X. and Sun, H., 2015. Analysis on the risk factors of intracranial infection secondary to traumatic brain injury. *Chinese Journal of Traumatology, 18*(2), pp. 81–83.

Liu, D.X., He, X., Wu, D., Zhang, Q., Yang, C., Liang, F.Y., He, X.F., Dai, G.Y., Pei, Z., Lan, Y. and Xu, G.Q., 2017. Continuous theta burst stimulation facilitates the clearance efficiency of the glymphatic pathway in a mouse model of sleep deprivation. *Neuroscience letters, 653*, pp. 189–194.

Liu, R.T., Walsh, R.F. and Sheehan, A.E., 2019. Prebiotics and probiotics for depression and anxiety: a systematic review and meta-analysis of controlled clinical trials. *Neuroscience & Biobehavioral Reviews, 102*, pp. 13–23.

Liu, Y., Zhu, C., Guo, J., Chen, Y. and Meng, C., 2020. The Neuroprotective Effect of Irisin in Ischemic Stroke. *Frontiers in Aging Neuroscience, 12*, p. 475.

Locher, R.J., Lünnemann, T., Garbe, A., Schaser, K.D., Schmidt-Bleek, K., Duda, G. and Tsitsilonis, S., 2015. Traumatic brain injury and bone healing: radiographic and biomechanical analyses of bone formation and stability in a combined murine trauma model. *Journal of musculoskeletal & neuronal interactions, 15*(4), p. 309.

Lowe, A., Bailey, M., O'Shaughnessy, T. and Macavei, V., 2020. Treatment of sleep disturbance following stroke and traumatic brain injury: a systematic review of conservative interventions. *Disability and Rehabilitation*, pp. 1–13.

Lowe, A., Neligan, A. and Greenwood, R., 2020. Sleep disturbance and recovery during rehabilitation after traumatic brain injury: a systematic review. *Disability and rehabilitation, 42*(8), pp. 1041–1054.

Lubetkin, E.I. and Jia, H., 2018. Burden of disease due to sleep duration and sleep problems in the elderly. *Sleep health, 4*(2), pp. 182–187.

Lucassen, E.A., De Mutsert, R., Le Cessie, S., Appelman-Dijkstra, N.M., Rosendaal, F.R., Van Heemst, D., Den Heijer, M., Biermasz, N.R. and NEO Study Group, 2017. Poor sleep quality and later sleep timing are risk factors for osteopenia and sarcopenia in middle-aged men and women: The NEO study. *PloS one, 12*(5), p. e0176685.

Ludwig, R., Vaduvathiriyan, P. and Siengsukon, C., 2020. Does cognitive-behavioural therapy improve sleep outcomes in individuals with traumatic brain injury: a scoping review. *Brain Injury, 34*(12), pp. 1569–1578.

Lugaresi, E., Provini, F. and Montagna, P., 2021. The neuroanatomy of sleep. Considerations on the role of the thalamus in sleep and a proposal for a caudorostral organization. *European Journal of Anatomy, 8*(2), pp. 85–93.

Ma, J., Qiao, P., Li, Q., Wang, Y., Zhang, L., Yan, L.J. and Cai, Z., 2019. Vagus nerve stimulation as a promising adjunctive treatment for ischemic stroke. *Neurochemistry international, 131*, p. 104539.

Makley, M.J., English, J.B., Drubach, D.A., Kreuz, A.J., Celnik, P.A. and Tarwater, P.M., 2008. Prevalence of sleep disturbance in closed head injury patients in a rehabilitation unit. *Neurorehabilitation and neural repair, 22*(4), pp. 341–347.

Maestroni, G.J., 2001. The immunotherapeutic potential of melatonin. *Expert opinion on investigational drugs, 10*(3), pp. 467–476.

Malhi, G.S., Adams, D., Porter, R., Wignall, A., Lampe, L., O'connor, N., Paton, M., Newton, L.A., Walter, G., Taylor, A. and Berk, M., 2009. Clinical practice recommendations for depression. *Acta Psychiatrica Scandinavica, 119*, pp. 8–26.

Mahoney, C.E., Mochizuki, T. and Scammell, T.E., 2020. Dual orexin receptor antagonists increase sleep and cataplexy in wild type mice. *Sleep*, *43*(6), p. zsz302.

Majid, M.S., Ahmad, H.S., Bizhan, H., Hosein, H.Z.M. and Mohammad, A., 2018. The effect of vitamin D supplement on the score and quality of sleep in 20–50 year old people with sleep disorders compared with control group. *Nutritional neuroscience*, *21*(7), pp. 511–519.

Makin, S.D.J., Turpin, S., Dennis, M.S. and Wardlaw, J.M., 2013. Cognitive impairment after lacunar stroke: systematic review and meta-analysis of incidence, prevalence and comparison with other stroke subtypes. *Journal of Neurology, Neurosurgery & Psychiatry*, *84*(8), pp. 893–900.

Malhi, G.S., Bell, E., Singh, A., Bassett, D., Berk, M., Boyce, P., Bryant, R., Gitlin, M., Hamilton, A., Hazell, P. and Hopwood, M., 2020. The 2020 royal Australian and New Zealand college of psychiatrists clinical practice guidelines for mood disorders: major depression summary. *Bipolar Disorders*.

Mancini, C.B., 2017. Decreasing Cost in the GI Endoscopy Suite by Utilizing Best Sedation Practices. **65**. Doctoral Projects. University of Southern Mississippi.

Mansournia, M.A., Ostadmohammadi, V., Doosti-Irani, A., Ghayour-Mobarhan, M., Ferns, G., Akbari, H., Ghaderi, A., Talari, H.R. and Asemi, Z., 2018. The effects of vitamin D supplementation on biomarkers of inflammation and oxidative stress in diabetic patients: a systematic review and meta-analysis of randomized controlled trials. *Hormone and Metabolic Research*, *50*(06), pp. 429–440.

Mari, A., Abu Baker, F., Mahamid, M., Sbeit, W. and Khoury, T., 2020. The evolving role of gut microbiota in the management of irritable bowel syndrome: An overview of the current knowledge. *Journal of clinical medicine*, *9*(3), p. 685.

Marin, J.M., Carrizo, S.J., Vicente, E. and Agusti, A.G., 2005. Long-term cardiovascular outcomes in men with obstructive sleep apnoea-hypopnoea with or without treatment with continuous positive airway pressure: an observational study. *The Lancet*, *365*(9464), pp. 1046–1053.

Marion, C.M., Radomski, K.L., Cramer, N.P., Galdzicki, Z. and Armstrong, R.C., 2018. Experimental traumatic brain injury identifies distinct early and late phase axonal conduction deficits of white matter pathophysiology, and reveals intervening recovery. *Journal of Neuroscience*, *38*(41), pp. 8723–8736.

Marón, F.J.M., Ferder, L., Reiter, R.J. and Manucha, W., 2020. Daily and seasonal mitochondrial protection: Unraveling common possible mechanisms involving vitamin D and melatonin. *The Journal of Steroid Biochemistry and Molecular Biology*, *199*. May.

Marseglia, L., D'Angelo, G., Manti, S., Rulli, I., Salvo, V., Buonocore, G., Reiter, R.J. and Gitto, E., 2017. Melatonin secretion is increased in children with severe traumatic brain injury. *International journal of molecular sciences*, *18*(5), p. 1053.

Marsh, H., 2014. *Do no harm: stories of life, death and brain surgery*. UK: Weidenfeld & Nicolson.

Martindale, S.L., Farrell-Carnahan, L.V., Ulmer, C.S., Kimbrel, N.A., McDonald, S.D. and Rowland, J.A., 2017. Sleep quality in returning veterans: The influence of mild traumatic brain injury. *Rehabilitation psychology*, *62*(4), p. 563.

Masi, G., Fantozzi, P., Villafranca, A., Tacchi, A., Ricci, F., Ruglioni, L., Inguaggiato, E., Pfanner, C. and Cortese, S., 2019. Effects of melatonin in children with attention-deficit/hyperactivity disorder with sleep disorders after methylphenidate treatment. *Neuropsychiatric disease and treatment*, *15*, p. 663.

Matthews, L.R., Danner, O.K., Ahmed, Y.A., Dennis-Griggs, D.M., Frederick, A., Clark, C., Moore, R., DuMornay, W., Childs, E.W. and Wilson, K.L., 2013. Combination therapy with vitamin D3, progesterone, omega-3 fatty acids, and glutamine reverses coma and improves clinical outcomes in patients with severe traumatic brain injuries: a case series. *Int J Case Rep Images*, *4*(3), pp. 143–148.

Maurer, L.F., Espie, C.A. and Kyle, S.D., 2018. How does sleep restriction therapy for insomnia work? A systematic review of mechanistic evidence and the introduction of the Triple-R model. *Sleep medicine reviews*, *42*, pp. 127–138.

Mayford, M., Siegelbaum, S.A. and Kandel, E.R., 2012. Synapses and memory storage. *Cold Spring Harbor perspectives in biology*, *4*(6), p. a005751.

McCarley, R.W. and Hobson, J.A., 1975. Neuronal excitability modulation over the sleep cycle: a structural and mathematical model. *Science*, *189*(4196), pp. 58–60.

McCarty, D.E., 2010. Resolution of hypersomnia following identification and treatment of vitamin D deficiency. *Journal of Clinical Sleep Medicine*, *6*(6), pp. 605–608.

McFarland, L.V., 2020. Efficacy of single-strain probiotics versus multi-strain mixtures: systematic review of strain and disease specificity. *Digestive diseases and sciences*, *66*, pp. 694–704.

McFarland, L.V. and Dublin, S., 2008. Meta-analysis of probiotics for the treatment of irritable bowel syndrome. *World journal of gastroenterology: WJG*, *14*(17), p. 2650.

McGinn, M.J. and Povlishock, J.T., 2016. Pathophysiology of traumatic brain injury. *Neurosurgery Clinics*, *27*(4), pp. 397–407.

McGinty, D.J. and Sterman, M.B., 1968. Sleep suppression after basal forebrain lesions in the cat. *Science*, *160*(3833), pp. 1253–1255.

Meldrum, B.S., 2000. Glutamate as a neurotransmitter in the brain: review of physiology and pathology. *The Journal of nutrition*, *130*(4), pp. 1007S–1015S.

Meltzer, L.J., Short, M., Booster, G.D., Gradisar, M., Marco, C.A., Wolfson, A.R. and Carskadon, M.A., 2019. Pediatric motor activity during sleep as measured by actigraphy. *Sleep*, *42*(1), p. zsy196.

Menn, S.J., Yang, R. and Lankford, A., 2014. Armodafinil for the treatment of excessive sleepiness associated with mild or moderate closed traumatic brain injury: a 12-week, randomized, double-blind study followed by a 12-month open-label extension. *Journal of Clinical Sleep Medicine*, *10*(11), pp. 1181–1191.

Menda, G., Bar, H.Y., Arthur, B.J., Rivlin, P.K., Wyttenbach, R.A., Strawderman, R.L. and Hoy, R.R., 2011. Classical conditioning through auditory stimuli in Drosophila: methods and models. *Journal of Experimental Biology*, *214*(17), pp. 2864–2870.

Menon, B.K., Saver, J.L., Prabhakaran, S., Reeves, M., Liang, L., Olson, D.M., Peterson, E.D., Hernandez, A.F., Fonarow, G.C., Schwamm, L.H. and Smith, E.E., 2012. Risk score for intracranial hemorrhage in patients with acute ischemic stroke treated with intravenous tissue-type plasminogen activator. *Stroke*, *43*(9), pp. 2293–2299.

Mestre H., Mori Y., Nedergaard M., 2020. The Brain's Glymphatic System: Current Controversies. *Trends in Neuroscience*, Jul; *43*(7): 458–466.

Miao, S., Zhao, C., Zhu, J., Hu, J., Dong, X. and Sun, L., 2018. Dietary soybean meal affects intestinal homoeostasis by altering the microbiota, morphology and inflammatory cytokine gene expression in northern snakehead. *Scientific reports*, *8*(1), pp. 1–10

Miao, Z., Schultzberg, M., Wang, X., & Zhao, Y., 2021. Role of polyunsaturated fatty acids in ischemic stroke – A perspective of specialized pro-resolving mediators. *Clinical Nutrition*, *40*(5), pp. 2974–2987.

Mieda, M. and Sakurai, T., 2016. Orexin (hypocretin) and narcolepsy. In *Narcolepsy* (pp. 11–23). Cham: Springer.

Mien, I.H., Chua, E.C.P., Lau, P., Tan, L.C., Lee, I.T.G., Yeo, S.C., Tan, S.S. and Gooley, J.J., 2014. Effects of exposure to intermittent versus continuous red light on human circadian rhythms, melatonin suppression, and pupillary constriction. *PloS one*, *9*(5), p. e96532.

Miller, C.B., Espie, C.A., Epstein, D.R., Friedman, L., Morin, C.M., Pigeon, W.R., Spielman, A.J. and Kyle, S.D., 2014. The evidence base of sleep restriction therapy for treating insomnia disorder. *Sleep medicine reviews*, *18*(5), pp. 415–424.

Min, J.Y. and Min, K.B., 2018. Outdoor artificial nighttime light and use of hypnotic medications in older adults: a population-based cohort study. *Journal of Clinical Sleep Medicine*, *14*(11), pp. 1903–1910.

Minen, M., Jinich, S. and Vallespir Ellett, G., 2019. Behavioral Therapies and Mind-Body Interventions for Posttraumatic Headache and Post-Concussive Symptoms: A Systematic Review. *Headache: The Journal of Head and Face Pain*, *59*(2), pp. 151–163.

Mischoulon, D., 2018. Popular herbal and natural remedies used in psychiatry. *Focus*, *16*(1), pp. 2–11.

Mitchell, A.J., Sheth, B., Gill, J., Yadegarfar, M., Stubbs, B., Yadegarfar, M. and Meader, N., 2017. Prevalence and predictors of post-stroke mood disorders: A meta-analysis and meta-regression of depression, anxiety and adjustment disorder. *General Hospital Psychiatry*, *47*, pp. 48–60.

Montoya, J.G., Holmes, T.H., Anderson, J.N., Maecker, H.T., Rosenberg-Hasson, Y., Valencia, I.J., Chu, L., Younger, J.W., Tato, C.M. and Davis, M.M., 2017. Cytokine signature associated with disease severity in chronic fatigue syndrome patients. *Proceedings of the National Academy of Sciences*, *114*(34), pp. E7150–E7158.

Momjian-Mayor, I. and Baron, J.C., 2005. The pathophysiology of watershed infarction in internal carotid artery disease: review of cerebral perfusion studies. *Stroke*, *36*(3), pp. 567–577.

Moruzzi, G. and Magoun, H.W., 1949. Brain stem reticular formation and activation of the EEG. *Electroencephalography and clinical neurophysiology*, *1*(1–4), pp. 455–473.

Mracsko, E. and Veltkamp, R., 2014. Neuroinflammation after intracerebral hemorrhage. *Frontiers in cellular neuroscience*, *8*, p. 388.

Mueller, C., Lin, J.C., Sheriff, S., Maudsley, A.A. and Younger, J.W., 2020. Evidence of widespread metabolite abnormalities in Myalgic encephalomyelitis/chronic fatigue syndrome: assessment with whole-brain magnetic resonance spectroscopy. *Brain imaging and behavior*, *14*(2), pp. 562–572.

Mullington, J.M., Simpson, N.S., Meier-Ewert, H.K. and Haack, M., 2010. Sleep loss and inflammation. *Best practice & research Clinical endocrinology & metabolism*, *24*(5), pp. 775–784.

Murphy, M., Riedner, B.A., Huber, R., Massimini, M., Ferrarelli, F. and Tononi, G., 2009. Source modeling sleep slow waves. *Proceedings of the National Academy of Sciences*, *106*(5), pp. 1608–1613.

Muscogiuri, G., Barrea, L., Scannapieco, M., Di Somma, C., Scacchi, M., Aimaretti, G., Savastano, S., Colao, A. and Marzullo, P., 2019. The lullaby of the sun: the role of vitamin D in sleep disturbance. *Sleep medicine*, *54*, pp. 262–265.

Nakase-Richardson, R., Holcomb, E.M., Schwartz, D.J., McCarthy, M., Thomas, B. and Barnett, S.D., 2016. Incidence, characterization, and predictors of sleep apnea in consecutive brain injury rehabilitation admissions. *Journal of Head Trauma Rehabilitation*, *31*(2), pp. 82–100.

Nakase-Richardson, R., Sherer, M., Barnett, S.D., Yablon, S.A., Evans, C.C., Kretzmer, T., Schwartz, D.J. and Modarres, M., 2013. Prospective evaluation of the nature, course, and impact of acute sleep abnormality after traumatic brain injury. *Archives of Physical Medicine and Rehabilitation*, *94*(5), pp. 875–882.

Nampiaparampil, D.E., 2008. Prevalence of chronic pain after traumatic brain injury: a systematic review. *Jama*, *300*(6), pp. 711–719.

Nauta, W., 1946. Hypothalamic regulation of sleep in rats J. *Neurophysiology*, *9*, pp. 283–316.

Navarro-Orozco, D. and Sánchez-Manso, J.C., 2020. Neuroanatomy, middle cerebral artery. [Updated 2020 Aug 22]. In *StatPearls [Internet]*. Treasure Island (FL): StatPearls Publishing. Available from: https://www.ncbi.nlm.nih.gov/books/NBK526002/

Nakazato, Y., Yoshimaru, K., Ohkuma, A., Araki, N., Tamura, N. and Shimazu, K., 2004. Central post-stroke pain in Wallenberg syndrome. *Brain and nerve*, *56*(5), pp. 385–388.

Neale, G. and Smith, A.J., 2007. Self-harm and suicide associated with benzodiazepine usage. *British journal of general practice*, *57*(538), pp. 407–408.

Neren, D., Johnson, M.D., Legon, W., Bachour, S.P., Ling, G. and Divani, A.A., 2016. Vagus nerve stimulation and other neuromodulation methods for treatment of traumatic brain injury. *Neurocritical care*, *24*(2), pp. 308–319.

Ng, S.Y. and Lee, A.Y.W., 2019. Traumatic brain injuries: pathophysiology and potential therapeutic targets. *Frontiers in cellular neuroscience*, *13*, p. 528.

Nguyen, S., McKay, A., Wong, D., Rajaratnam, S.M., Spitz, G., Williams, G., Mansfield, D. and Ponsford, J.L., 2017. Cognitive behavior therapy to treat sleep disturbance and fatigue after traumatic brain injury: a pilot randomized controlled trial. *Archives of physical medicine and rehabilitation*, *98*(8), pp. 1508–1517.

Nielsen, N. and Friberg, H., 2013. Can we conclude anything yet about the effect of hypothermia for patients arresting in-hospital? *Resuscitation*, *84*(5), pp. 535–536.

NIH Monograph. Valerian – Health Professional Fact Sheet. Ods.od.nih.gov. 2021. last updated 2013.

Noh, H.J., Joo, E.Y., Kim, S.T., Yoon, S.M., Koo, D.L., Kim, D., Lee, G.H. and Hong, S.B., 2012. The relationship between hippocampal volume and cognition in patients with chronic primary insomnia. *Journal of clinical neurology (Seoul, Korea)*, *8*(2), p. 130.

Nolan, J.P., Neumar, R.W., Adrie, C., Aibiki, M., Berg, R.A., Böttiger, B.W., Callaway, C., Clark, R.S., Geocadin, R.G., Jauch, E.C. and Kern, K.B., 2008. Post-cardiac arrest syndrome: epidemiology, pathophysiology, treatment, and prognostication: a scientific statement from the International liaison Committee on Resuscitation; the American Heart Association Emergency Cardiovascular Care Committee; the Council on Cardiovascular Surgery and Anesthesia; the Council on cardiopulmonary, Perioperative, and Critical Care; the Council on clinical cardiology; the Council on Stroke. *Resuscitation*, *79*(3), pp. 350–379.

Noonan, K., Crewther, S.G., Carey, L.M., Pascoe, M.C. and Linden, T., 2013. Sustained inflammation 1.5 years post-stroke is not associated with depression in elderly stroke survivors. *Clinical interventions in aging*, *8*, p. 69.

Nouh, A., Remke, J. and Ruland, S., 2014. Ischemic posterior circulation stroke: a review of anatomy, clinical presentations, diagnosis, and current management. *Frontiers in neurology*, *5*, p. 30.

Nuwer, M.R., Hovda, D.A., Schrader, L.M. and Vespa, P.M., 2005. Routine and quantitative EEG in mild traumatic brain injury. *Clinical Neurophysiology*, *116*(9), pp. 2001–2025.

O'Callaghan, F., Muurlink, O. and Reid, N., 2018. Effects of caffeine on sleep quality and daytime functioning. *Risk management and healthcare policy*, *11*, p. 263.

Ochs-Balcom, H.M., Hovey, K.M., Andrews, C., Cauley, J.A., Hale, L., Li, W., Bea, J.W., Sarto, G.E., Stefanick, M.L., Stone, K.L. and Watts, N.B., 2020. Short sleep is associated with low bone mineral density and osteoporosis in the Women's Health Initiative. *Journal of Bone and Mineral Research*, *35*(2), pp. 261–268.

Office for National Statistics, 2017. *Cancer Registration Statistics*. [online] England: National Statistics, p. 2.

Oh, Y.H., Kim, H., Kong, M., Oh, B. and Moon, J.H., 2019. Association between weekend catch-up sleep and health-related quality of life of Korean adults. *Medicine*, *98*(13). e14966. doi: 10.1097/MD.0000000000014966.

Oldham, M.A. and Ciraulo, D.A., 2014. Bright light therapy for depression: a review of its effects on chronobiology and the autonomic nervous system. *Chronobiology international*, *31*(3), pp. 305–319.

Olofsson, P.S., Levine, Y.A., Caravaca, A., Chavan, S.S., Pavlov, V.A., Faltys, M. and Tracey, K.J., 2015. Single-pulse and unidirectional electrical activation of the cervical vagus nerve reduces tumor necrosis factor in endotoxemia. *Bioelectronic Medicine*, *2*(1), pp. 37–42.

Onega, L.L. and Pierce, T.W., 2020. Use of bright light therapy for older adults with dementia. *BJPsych Advances*, *26*(4), pp. 221–228.

Osborn, A.J., Mathias, J.L. and Fairweather-Schmidt, A.K., 2016. Prevalence of anxiety following adult traumatic brain injury: A meta-analysis comparing measures, samples and postinjury intervals. *Neuropsychology*, *30*(2), p. 247.

Ouellet, M.C., Beaulieu-Bonneau, S. and Morin, C.M., 2015. Sleep-wake disturbances after traumatic brain injury. *The Lancet Neurology*, *14*(7), pp. 746–757.

Ouellet, M.C. and Morin, C.M., 2006. Subjective and objective measures of insomnia in the context of traumatic brain injury: a preliminary study. *Sleep medicine*, *7*(6), pp. 486–497.

Ouwehand, A.C., Invernici, M.M., Furlaneto, F.A. and Messora, M.R., 2018. Effectiveness of multi-strain versus single-strain probiotics: current status and recommendations for the future. *Journal of clinical gastroenterology*, *52*, pp. S35–S40.

Owens, J.A., Spitz, G., Ponsford, J.L., Dymowski, A.R. and Willmott, C., 2018. An investigation of white matter integrity and attention deficits following traumatic brain injury. *Brain injury*, *32*(6), pp. 776–783.

Palagini, L., Baglioni, C., Ciapparelli, A., Gemignani, A. and Riemann, D., 2013. REM sleep dysregulation in depression: state of the art. *Sleep medicine reviews*, *17*(5), pp. 377–390.

Palomäki, H., Berg, A., Meririnne, E., Kaste, M., Lönnqvist, R., Lehtihalmes, M. and Lönnqvist, J., 2003. Complaints of poststroke insomnia and its treatment with mianserin. *Cerebrovascular Diseases*, *15*(1–2), pp. 56–62.

Parcell, D.L., Ponsford, J.L., Redman, J.R. and Rajaratnam, S.M., 2008. Poor sleep quality and changes in objectively recorded sleep after traumatic brain injury: a preliminary study. *Archives of physical medicine and rehabilitation*, *89*(5), pp. 843–850.

Parikh, O.A., 2017. A Comparison of Moderate Oral Sedation Drug Regimens for Pediatric Dental Treatment: A Pilot Study.

Park, C.Y., 2018. Night light pollution and ocular fatigue. *Journal of Korean medical science*, *33*(38), e257. https://doi.org/10.3346/jkms.2018.33.e257

Parvizi, J. and Damasio, A.R., 2003. Neuroanatomical correlates of brainstem coma. *Brain*, *126*(7), pp. 1524–1536.

Pascoe, M.C., Skoog, I., Blomstrand, C. and Linden, T., 2015. Albumin and depression in elderly stroke survivors: An observational cohort study. *Psychiatry research*, *230*(2), pp. 658–663.

Pascoe, M.C., Thompson, D.R., Jenkins, Z.M. and Ski, C.F., 2017. Mindfulness mediates the physiological markers of stress: systematic review and meta-analysis. *Journal of psychiatric research*, *95*, pp. 156–178

Pathare, N., Sushilkumar, S., Haley, L., Jain, S. and Osier, N., 2020. The impact of traumatic brain injury on microbiome composition: a systematic review. *Biological research for nursing*, *22*(4), pp. 495–505.

Peng, W., Wu, Z., Song, K., Zhang, S., Li, Y. and Xu, M., 2020. Regulation of sleep homeostasis mediator adenosine by basal forebrain glutamatergic neurons. *Science*, *369*(6508), eabb0556. doi: 10.1126/science.abb0556.

Perkins, G.D. and Brace-McDonnell, S.J., 2015. The UK out of hospital cardiac arrest outcome (OHCAO) project. *BMJ open*, *5*(10).

Perlis, M.L., Artiola, L. and Giles, D.E., 1997. Sleep complaints in chronic post-concussion syndrome. *Perceptual and motor skills*, *84*(2), pp. 595–599.

Perron, S., Tétreault, L.F., King, N., Plante, C. and Smargiassi, A., 2012. Review of the effect of aircraft noise on sleep disturbance in adults. *Noise and health*, *14*(57), p. 58.

Petursson, H., 1994. The benzodiazepine withdrawal syndrome. *Addiction*, *89*(11), pp. 1455–1459.

Piantoni, G., Halgren, E. and Cash, S.S., 2017. Spatiotemporal characteristics of sleep spindles depend on cortical location. *Neuroimage*, *146*, pp. 236–245.

Piovezan, R.D., Hirotsu, C., Feres, M.C., Cintra, F.D., Andersen, M.L., Tufik, S. and Poyares, D., 2017. Obstructive sleep apnea and objective short sleep duration are independently associated with the risk of serum vitamin D deficiency. *PloS one*, *12*(7), p. e0180901.

Pirrera, S., De Valck, E. and Cluydts, R., 2010. Nocturnal road traffic noise: A review on its assessment and consequences on sleep and health. *Environment international*, *36*(5), pp. 492–498.

Plaza-Díaz J., Ruiz-Ojeda F.J., Vilchez-Padial L.M., 2017. Gil A. Evidence of the Anti-Inflammatory Effects of Probiotics and Synbiotics in Intestinal Chronic Diseases. *Nutrients*. 9(6):555.

Poh, J.H. and Chee, M.W., 2017. Degradation of cortical representations during encoding following sleep deprivation. *Neuroimage*, *153*, pp. 131–138.

Ponsford, J.L., Parcell, D.L., Sinclair, K.L., Roper, M. and Rajaratnam, S.M., 2013. Changes in sleep patterns following traumatic brain injury: a controlled study. *Neurorehabilitation and neural repair*, *27*(7), pp. 613–621.

Porter, B.A., Khodaparast, N., Fayyaz, T., Cheung, R.J., Ahmed, S.S., Vrana, W.A., Rennaker, R.L. and Kilgard, M.P., 2012. Repeatedly pairing vagus nerve stimulation with a movement reorganizes primary motor cortex. *Cerebral Cortex*, *22*(10), pp. 2365–2374.

Posadzki, P., Cramer, H., Kuzdzal, A., Lee, M.S. and Ernst, E., 2014. Yoga for hypertension: a systematic review of randomized clinical trials. *Complementary Therapies in Medicine*, *22*(3), pp. 511–522.

Postolache, T.T., Wadhawan, A., Can, A., Lowry, C.A., Woodbury, M., Makkar, H., Hoisington, A.J., Scott, A.J., Potocki, E., Benros, M.E. and Stiller, J.W., 2020. Inflammation in traumatic brain injury. *Journal of Alzheimer's Disease*, *74*(1), pp. 1–28.

Powell, M.R., Brown, A.W., Klunk, D., Geske, J.R., Krishnan, K., Green, C. and Bergquist, T.F., 2019. Injury severity and depressive symptoms in a post-acute brain injury rehabilitation sample. *Journal of clinical psychology in medical settings*, *26*(4), pp. 470–482.

Prakash, A., Harsh, V., Gupta, U., Kumar, J. and Kumar, A., 2018. Depressed fractures of skull: an institutional series of 453 patients and brief review of literature. *Asian journal of neurosurgery*, *13*(2), p. 222.

Prins, M.L., Alexander, D., Giza, C.C. and Hovda, D.A., 2013. Repeated mild traumatic brain injury: mechanisms of cerebral vulnerability. *Journal of neurotrauma*, *30*(1), pp. 30–38.

Prospero-García, O., Jiménez-Anguiano, A. and Drucker-Colín, R., 1993. The combination of VIP and atropine induces REM sleep in cats rendered insomniac by PCPA. *Neuropsychopharmacology*, *8*(4), pp. 387–390.

Prospéro-García, O., Méndez-Díaz, M., Ruiz-Contreras, A.E. and Pérez-Morales, M., 2011. Neuropeptides and REM sleep. *Rapid Eye Movement Sleep: Regulation and Function*, pp. 247–256.

Provencio, I., Rodriguez, I.R., Jiang, G., Hayes, W.P., Moreira, E.F. and Rollag, M.D., 2000. A novel human opsin in the inner retina. *Journal of Neuroscience*, *20*(2), pp. 600–605.

Purves, D., Augustine G.J., Fitzpatrick D., Hall W.C., Lamantia A.-S., McNamara J.O., Williams S.M., 2011. *Neuroscience*, 3rd Edition. Massachusetts: Sinauer Associates.

Puthanveettil, S. and Kandel, E., 2011. Molecular mechanisms for the initiation and maintenance of long-term memory storage. In Curran T. and Christen Y. (eds.), *Two Faces of Evil: Cancer and Neurodegeneration. Research and Perspectives in Alzheimer's Disease*, pp. 143–160. Berlin, Heidelberg: Springer. https://doi.org/10.1007/978-3-642-16602-0_13

Rahman, S.A., Wright, K.P., Lockley, S.W., Czeisler, C.A. and Gronfier, C., 2019. Characterizing the temporal dynamics of melatonin and cortisol changes in response to nocturnal light exposure. *Scientific Reports*, *9*(1), pp. 1–12.

Raikes, A.C., Dailey, N.S., Shane, B.R., Forbeck, B., Alkozei, A. and Killgore, W.D., 2020. Daily morning blue light therapy improves daytime sleepiness, sleep quality, and quality of life following a mild traumatic brain injury. *The Journal of Head Trauma Rehabilitation*, *35*(5), pp. E405–E421.

Raikes, A.C. and Schaefer, S.Y., 2016. Sleep quantity and quality during acute concussion: a pilot study. *Sleep*, *39* (12), pp. 2141–2147.

Rajkovic, O., Potjewyd, G. and Pinteaux, E., 2018. Regenerative medicine therapies for targeting neuroinflammation after stroke. *Frontiers in Neurology*, *9*, p. 734.

Ramlackhansingh, A.F., Brooks, D.J., Greenwood, R.J., Bose, S.K., Turkheimer, F.E., Kinnunen, K.M., Gentleman, S., Heckemann, R.A., Gunanayagam, K., Gelosa, G. and Sharp, D.J., 2011. Inflammation after trauma: microglial activation and traumatic brain injury. *Annals of neurology*, *70*(3), pp. 374–383.

Rammohan, K.W., Rosenberg, J.H., Lynn, D.J., Blumenfeld, A.M., Pollak, C.P. and Nagaraja, H.N., 2002. Efficacy and safety of modafinil (Provigil®) for the treatment of fatigue in multiple sclerosis: a two centre phase 2 study. *Journal of Neurology, Neurosurgery & Psychiatry*, *72*(2), pp. 179–183.

Rankhambe, H.B. and Pande, S., 2020. Effect of "Om" chanting on anxiety in bus drivers. *National Journal of Physiology, Pharmacy and Pharmacology*, *10*(12), pp. 1138–1141.

Rao, M.L., Gross, G., Strebel, B., Halaris, A., Huber, G., Bräunig, P. and Marler, M., 1994. Circadian rhythm of tryptophan, serotonin, melatonin, and pituitary hormones in schizophrenia. *Biological Psychiatry*, *35*(3), pp. 151–163.

Rao, N.P., Deshpande, G., Gangadhar, K.B., Arasappa, R., Varambally, S., Venkatasubramanian, G. and Ganagadhar, B.N., 2018. Directional brain networks underlying OM chanting. *Asian Journal of Psychiatry*, *37*, pp. 20–25.

Rauchs, G., Schabus, M., Parapatics, S., Bertran, F., Clochon, P., Hot, P., Denise, P., Desgranges, B., Eustache, F., Gruber, G. and Anderer, P., 2008. Is there a link between sleep changes and memory in Alzheimer's disease? *Neuroreport*, *19*(11), p. 1159.

Rhea, E.M., Logsdon, A.F., Hansen, K.M., Williams, L.M., Reed, M.J., Baumann, K.K., Holden, S.J., Raber, J., Banks, W.A. and Erickson, M.A., 2021. The S1 protein of SARS-CoV-2 crosses the blood–brain barrier in mice. *Nature Neuroscience*, *24*(3), pp. 368–378.

Richardson, K., Fox, C., Maidment, I., Steel, N., Loke, Y.K., Arthur, A., Myint, P.K., Grossi, C.M., Mattishent, K., Bennett, K. and Campbell, N.L., 2018. Anticholinergic drugs and risk of dementia: case-control study. *BMJ*, *361*, k1315.

Riemann, D., Baglioni, C., Bassetti, C., Bjorvatn, B., Dolenc Groselj, L., Ellis, J.G., Espie, C.A., Garcia-Borreguero, D., Gjerstad, M., Gonçalves, M. and Hertenstein, E., 2017. European guideline for the diagnosis and treatment of insomnia. *Journal of Sleep Research*, *26*(6), pp. 675–700.

Rios, P., Cardoso, R., Morra, D., Nincic, V., Goodarzi, Z., Farah, B., Harricharan, S., Morin, C.M., Leech, J., Straus, S.E. and Tricco, A.C., 2019. Comparative effectiveness and safety of pharmacological and non-pharmacological interventions for insomnia: an overview of reviews. *Systematic Reviews*, *8*(1), pp. 1–16.

Riva, M.A., Cimino, V. and Sanchirico, S., 2017. Gian Lorenzo Bernini's 17th century white noise machine. *The Lancet Neurology*, *16*(10), p. 776.

Rola, R., Wierzbicka, A., Wichniak, A., Jernajczyk, W., Richter, P. and Ryglewicz, D., 2007. Sleep related breathing disorders in patients with ischemic stroke and transient ischemic attacks: respiratory and clinical correlation. *Journal of Physiology and Pharmacology*, *58*(5), pp. 575–582.

Rosenberg, S., 2017. *Accessing the healing power of the vagus nerve: self-help exercises for anxiety, depression, trauma, and autism.* North Atlantic Books.

Roth, T., 2007. Insomnia: definition, prevalence, etiology, and consequences. *Journal of Clinical Sleep Medicine*, *3*(5 suppl), pp. S7–S10.

Röther, J., Ford, G.A. and Thijs, V.N., 2013. Thrombolytics in acute ischaemic stroke: historical perspective and future opportunities. *Cerebrovascular Diseases*, *35*(4), pp. 313–319.

Rye, D.B., Bliwise, D.L., Parker, K., Trotti, L.M., Saini, P., Fairley, J., Freeman, A., Garcia, P.S., Owens, M.J., Ritchie, J.C. and Jenkins, A., 2012. Modulation of vigilance in the primary hypersomnias by endogenous enhancement of GABAA receptors. *Science Translational Medicine*, *4*(161), pp. 161ra151–161ra151.

Ryttersgaard, T.O., Johnsen, S.P., Riis, J.Ø., Mogensen, P.H. and Bjarkam, C.R., 2020. Prevalence of depression after moderate to severe traumatic brain injury among adolescents and young adults: a systematic review. *Scandinavian Journal of Psychology*, *61*(2), pp. 297–306.

Sacks, Oliver (17 September 2007). The abyss. *The New Yorker.*

Said, S.I. and Mutt, V., 1970. Potent peripheral and splanchnic vasodilator peptide from normal gut. *Nature*, *225*(5235), pp. 863–864.

Salahudeen, M.S., Duffull, S.B. and Nishtala, P.S., 2015. Anticholinergic burden quantified by anticholinergic risk scales and adverse outcomes in older people: a systematic review. *BMC geriatrics*, *15*(1), pp. 1–14.

Saletin, J.M., Goldstein-Piekarski, A.N., Greer, S.M., Stark, S., Stark, C.E. and Walker, M.P., 2016. Human hippocampal structure: a novel biomarker predicting mnemonic vulnerability to, and recovery from, sleep deprivation. *Journal of Neuroscience*, *36*(8), pp. 2355–2363.

Salmond, C.H., Chatfield, D.A., Menon, D.K., Pickard, J.D. and Sahakian, B.J., 2005. Cognitive sequelae of head injury: involvement of basal forebrain and associated structures. *Brain*, *128*(1), pp. 189–200.

Sanchez, E., El-Khatib, H., Arbour, C., Bedetti, C., Blais, H., Marcotte, K., Baril, A.A., Descoteaux, M., Gilbert, D., Carrier, J. and Gosselin, N., 2019. Brain white matter damage and its association with neuronal synchrony during sleep. *Brain*, *142*(3), pp. 674–687.

Sanchez-Barcelo, E.J., Mediavilla, M.D., Tan, D.X. and Reiter, R.J., 2010. Clinical uses of melatonin: evaluation of human trials. *Current Medicinal Chemistry*, *17*(19), pp. 2070–2095.

Schabus, M., Dang-Vu, T.T., Albouy, G., Balteau, E., Boly, M., Carrier, J., Darsaud, A., Degueldre, C., Desseilles, M., Gais, S. and Phillips, C., 2007. Hemodynamic cerebral correlates of sleep spindles during human non-rapid eye movement sleep. *Proceedings of the National Academy of Sciences*, *104*(32), pp. 13164–13169.

Scoville, W.B., Milner, B., 1957. Loss of recent memory after bilateral hippocampal lesions. *Journal of Neurology, Neurosurgery & Psychiatry*, *20*(1), pp. 11–21.

Sedlmeier, P., Eberth, J., Schwarz, M., Zimmermann, D., Haarig, F., Jaeger, S. and Kunze, S., 2012. The psychological effects of meditation: a meta-analysis. *Psychological Bulletin*, *138*(6), p. 1139.

Sekhon, M.S., Ainslie, P.N. and Griesdale, D.E., 2017. Clinical pathophysiology of hypoxic ischemic brain injury after cardiac arrest: a "two-hit" model. *Critical Care*, *21*(1), pp. 1–10.

Sexton, E., Donnelly, N.A., Merriman, N., Guzman-Castillo, M., Bandosz, P., Wren, M.A., Hickey, A., O'Flaherty, M. and Bennett, K., 2019. OP51 Projecting the incidence and prevalence of post-stroke cognitive impairment and dementia in the irish population aged 40+ years from 2015–2025. *Journal of Epidemiology and Community Health*, *73*(Suppl 1), p. A25.

Shekleton, J.A., Parcell, D.L., Redman, J.R., Phipps-Nelson, J., Ponsford, J.L. and Rajaratnam, S.M., 2010. Sleep disturbance and melatonin levels following traumatic brain injury. *Neurology*, *74*(21), pp. 1732–1738.

Shemie, S.D. and Gardiner, D., 2018. Circulatory arrest, brain arrest and death determination. *Frontiers in Cardiovascular Medicine*, *5*, p. 15.

Sherer, M., Yablon, S.A. and Nakase-Richardson, R., 2009. Patterns of recovery of posttraumatic confusional state in neurorehabilitation admissions after traumatic brain injury. *Archives of Physical Medicine and Rehabilitation*, *90*(10), pp. 1749–1754.

Sherman, S.M., 2014. The function of metabotropic glutamate receptors in thalamus and cortex. *The Neuroscientist*, *20*(2), pp. 136–149.

Shukla, D., Mahadevan, A., Sastry, K.V. and Shankar, S.K., 2007. Pathology of post traumatic brainstem and hypothalamic injuries. *Clinical Neuropathology*, *26*(5), pp. 197–209.

Simón-Arceo, K., Ramírez-Salado, I. and Calvo, J.M., 2003. Long-lasting enhancement of rapid eye movement sleep and pontogeniculooccipital waves by vasoactive intestinal peptide microinjection into the amygdala temporal lobe. *Sleep*, *26*(3), pp. 259–264.

Sinclair, K.L., Ponsford, J.L., Taffe, J., Lockley, S.W. and Rajaratnam, S.M., 2014. Randomized controlled trial of light therapy for fatigue following traumatic brain injury. *Neurorehabilitation and Neural Repair*, *28*(4), pp. 303–313.

Singhal, R.K., Chauhan, J., Jatav, H.S., Rajput, V.D., Singh, G.S. and Bose, B., 2021. Artificial night light alters ecosystem services provided by biotic components. *Biologia Futura*, pp. 1–17.

Skinner, B.F., 1951. How to teach animals. *Scientific American*, *185*(6), pp. 26–29.

Sloane, P.D., Figueiro, M. and Cohen, L., 2008. Light as therapy for sleep disorders and depression in older adults. *Clinical geriatrics*, *16*(3), p. 25.

Slobounov, S., Sebastianelli, W. and Hallett, M., 2012. Residual brain dysfunction observed one year post-mild traumatic brain injury: combined EEG and balance study. *Clinical Neurophysiology*, *123*(9), pp. 1755–1761.

Smaers, J.B., Dechmann, D.K., Goswami, A., Soligo, C. and Safi, K., 2012. Comparative analyses of evolutionary rates reveal different pathways to encephalization in bats, carnivorans, and primates. *Proceedings of the National Academy of Sciences*, *109*(44), pp. 18006–18011.

Smith, D.C., Modglin, A.A., Roosevelt, R.W., Neese, S.L., Jensen, R.A., Browning, R.A. and Clough, R.W., 2005. Electrical stimulation of the vagus nerve enhances cognitive and motor recovery following moderate fluid percussion injury in the rat. *Journal of Neurotrauma*, *22*(12), pp. 1485–1502.

Sommerfeld, D.K. and Welmer, A.K., 2012. Pain following stroke, initially and at 3 and 18 months after stroke, and its association with other disabilities. *European Journal of Neurology*, *19*(10), pp. 1325–1330.

Song, S.O., He, K., Narla, R.R., Kang, H.G., Ryu, H.U. and Boyko, E.J., 2019. Metabolic consequences of obstructive sleep apnea especially pertaining to diabetes mellitus and insulin sensitivity. *Diabetes & Metabolism Journal*, *43*(2), p. 144.

Spira, J.L., Lathan, C.E., Bleiberg, J. and Tsao, J.W., 2014. The impact of multiple concussions on emotional distress, post-concussive symptoms, and neurocognitive functioning in active duty United States marines independent of combat exposure or emotional distress. *Journal of Neurotrauma*, *31*(22), pp. 1823–1834.

Stamova, B., Ander, B.P., Jickling, G., Hamade, F., Durocher, M., Zhan, X., Liu, D.Z., Cheng, X., Hull, H., Yee, A. and Ng, K., 2019. The intracerebral hemorrhage blood transcriptome in humans differs from the ischemic stroke and vascular risk factor control blood transcriptomes. *Journal of Cerebral Blood Flow & Metabolism*, *39*(9), pp. 1818–1835.

Staner, L., 2003. Sleep and anxiety disorders. *Dialogues in Clinical Neuroscience*, *5*(3), p. 249.

Sterman, M.B. and Clemente, C.D., 1962. Forebrain inhibitory mechanisms: sleep patterns induced by basal forebrain stimulation in the behaving cat. *Experimental Neurology*, *6*(2), pp. 103–117.

Stern, A.L. and Naidoo, N., 2015. Wake-active neurons across aging and neurodegeneration: a potential role for sleep disturbances in promoting disease. *Springerplus*, *4*(1), pp. 1–13.

Stoneman, J., (2021). Homeopaths have 'crossed the line' peddling 'dangerous' vaccine myths. *Telegraph Newspapers*. 24 January 2021.

Strausz, S., Havulinna, A.S., Tuomi, T., Bachour, A., Groop, L., Mäkitie, A., Koskinen, S., Salomaa, V., Palotie, A., Ripatti, S. and Palotie, T., 2018. Obstructive sleep apnoea and the risk for coronary heart disease and type 2 diabetes: a longitudinal population-based study in Finland. *BMJ open*, *8*(10), p. e022752.

Stroke Association., 2017. *The State of the Nation*. Stroke statistics – January 2017.

Suh, M., Choi-Kwon, S. and Kim, J.S., 2014. Sleep disturbances after cerebral infarction: role of depression and fatigue. *Journal of Stroke and Cerebrovascular Diseases*, *23*(7), pp. 1949–1955.

Sullivan, C.E. and McNamara, S.G., 1998. Sleep apnoea and snoring: potential links with vascular disease. *Thorax*, *53*(suppl 3), pp. S8–S11.

Sullivan, M.P., Torres, S.J., Mehta, S. and Ahn, J., 2013. Heterotopic ossification after central nervous system trauma: a current review. *Bone & Joint Research*, *2*(3), pp. 51–57.

Swett, C.P. and Hobson, J.A., 1968. The effects of posterior hypothalamic lesions on behavioral and electrographic manifestations of sleep and waking in cats. *Archives Italiennes de Biologie*, *106*(3), pp. 283–293.

Szymusiak, R. and McGinty, D., 1986. Sleep-related neuronal discharge in the basal forebrain of cats. *Brain Research*, *370*(1), pp. 82–92.

Taheri, S. and Mignot, E., 2002. The genetics of sleep disorders. *The Lancet Neurology*, *1*(4), pp. 242–250.

Takeshima, M., Utsumi, T., Aoki, Y., Wang, Z., Suzuki, M., Okajima, I., Watanabe, N., Watanabe, K. and Takaesu, Y., 2020. Efficacy and safety of bright light therapy for manic and depressive symptoms in patients with bipolar disorder: a systematic review and meta-analysis. *Psychiatry and Clinical Neurosciences*, *74*(4), pp. 247–256.

Tang, W.K., Wang, L., Wong, G.K.C., Ungvari, G.S., Yasuno, F., Tsoi, K.K. and Kim, J.S., 2020. Depression after subarachnoid hemorrhage: a systematic review. *Journal of Stroke*, *22*(1), p. 11.

Tarocco, A., Caroccia, N., Morciano, G., Wieckowski, M.R., Ancora, G., Garani, G. and Pinton, P., 2019. Melatonin as a master regulator of cell death and inflammation: molecular mechanisms and clinical implications for newborn care. *Cell Death & Disease*, *10*(4), pp. 1–12.

Taylor, R.A. and Sansing, L.H., 2013. Microglial responses after ischemic stroke and intracerebral hemorrhage. *Clinical and Developmental Immunology*, *2013*, 746068. https://doi.org/10.1155/2013/746068

Temkin, N.R., Anderson, G.D., Winn, H.R., Ellenbogen, R.G., Britz, G.W., Schuster, J., Lucas, T., Newell, D.W., Mansfield, P.N., Machamer, J.E. and Barber, J., 2007. Magnesium sulfate for neuroprotection after traumatic brain injury: a randomised controlled trial. *The Lancet Neurology*, *6*(1), pp. 29–38.

Temple, J.L., Bernard, C., Lipshultz, S.E., Czachor, J.D., Westphal, J.A. and Mestre, M.A., 2017. The safety of ingested caffeine: a comprehensive review. *Frontiers in Psychiatry*, *8*, p. 80.

Thayer, J.F. and Sternberg, E., 2006. Beyond heart rate variability: vagal regulation of allostatic systems. *Annals of the New York Academy of Sciences*, *1088*(1), pp. 361–372.

Thomas, R.L., Jiang, L., Adams, J.S., Xu, Z.Z., Shen, J., Janssen, S., Ackermann, G., Vanderschueren, D., Pauwels, S., Knight, R. and Orwoll, E.S., 2020. Vitamin D metabolites and the gut microbiome in older men. *Nature Communications*, *11*(1), pp. 1–10.

Tietjens, J.R., Claman, D., Kezirian, E.J., De Marco, T., Mirzayan, A., Sadroonri, B., Goldberg, A.N., Long, C., Gerstenfeld, E.P. and Yeghiazarians, Y., 2019. Obstructive sleep apnea in cardiovascular disease: a review of the literature and proposed multidisciplinary clinical management strategy. *Journal of the American Heart Association*, *8*(1), p. e010440.

Tilley, A. and Brown, S., 1992. Sleep deprivation. *Handbook of Human Performance*, *3*, pp. 237–238.

Toman E., Bishop J.R.B., Davies D.J., et al, 2017. Vitamin D deficiency in traumatic brain injury and its relationship with severity of injury and quality of life: a prospective, observational study. *Journal of Neurotrauma*, *34*(7), pp. 1448–1456.

Tonegawa, S., Liu, X., Ramirez, S. and Redondo, R., 2015. Memory engram cells have come of age. *Neuron*, *87*(5), pp. 918–931.

Torres, E.R., Strack, E.F., Fernandez, C.E., Tumey, T.A. and Hitchcock, M.E., 2015. Physical activity and white matter hyperintensities: a systematic review of quantitative studies. *Preventive Medicine Reports*, *2*, pp. 319–325.

Traeger, J., Hoffman, B., Misencik, J., Hoffer, A. and Makii, J., 2020. Pharmacologic treatment of neurobehavioral sequelae following traumatic brain injury. *Critical Care Nursing Quarterly*, *43*(2), pp. 172–190.

Traumatic Brain Injury & Concussion, Facts, Data and Statistics. CDC, 1999, 2014 data, Last update, March, 2019.

Treister, A.K., Hatch, M.N., Cramer, S.C. and Chang, E.Y., 2017. Demystifying poststroke pain: from etiology to treatment. *PM&R*, *9*(1), pp. 63–75.

Trotti, L.M. and Arnulf, I., 2020. Idiopathic hypersomnia and other hypersomnia syndromes. *Neurotherapeutics*, *18*, pp. 20–31. https://doi.org/10.1007/s13311-020-00919-1

Tsubono, Y., Fukao, A. and Hisamichi, S., 1993. Health practices and mortality in a rural Japanese population. *The Tohoku Journal of Experimental Medicine*, *171*(4), pp. 339–348.

Tsuyama, J., Nakamura, A., Ooboshi, H., Yoshimura, A. and Shichita, T., 2018, November. Pivotal role of innate myeloid cells in cerebral post-ischemic sterile inflammation. In *Seminars in Immunopathology* (**Vol. 40**, No. 6, pp. 523–538). Berlin Heidelberg: Springer.

Tsai, J.H., Yang, P., Chen, C.C., Chung, W., Tang, T.C., Wang, S.Y. and Liu, J.K., 2009. Zolpidem-induced amnesia and somnambulism: rare occurrences? *European Neuropsychopharmacology*, *19*(1), pp. 74–76.

Tucker, B., Aston, J., Dines, M., Caraman, E., Yacyshyn, M., McCarthy, M. and Olson, J.E., 2017. Early brain edema is a predictor of in-hospital mortality in traumatic brain injury. *The Journal of Emergency Medicine*, *53*(1), pp. 18–29.

Tulving, E., 1972. Episodic and semantic memory. In: Tulving, E., Donaldson, W. (eds.), *Organization of memory*. New York: Academic Press, pp. 381–402.

US Department of Health & Human Services, Centres for Disease Control and Prevention, 2016. *Health, United States, with chartbook on long-term trends in health*. Washington: National Centre for Health Statistics, p. 49.

Valko, P.O., Gavrilov, Y.V., Yamamoto, M., Noaín, D., Reddy, H., Haybaeck, J., Weis, S., Baumann, C.R. and Scammell, T.E., 2016. Damage to arousal-promoting brainstem neurons with traumatic brain injury. *Sleep*, *39*(6), pp. 1249–1252.

Van Asch, C.J., Luitse, M.J., Rinkel, G.J., van der Tweel, I., Algra, A. and Klijn, C.J., 2010. Incidence, case fatality, and functional outcome of intracerebral haemorrhage over time, according to age, sex, and ethnic origin: a systematic review and meta-analysis. *The Lancet Neurology*, *9*(2), pp. 167–176.

Varvogli, L. and Darvin, C., 2011. Stress management techniques: evidence-based procedures that reduce stress and promote health. *Health Science Journal*, *5*(2), 74.

van der Velden, A.M., Kuyken, W., Wattar, U., Crane, C., Pallesen, K.J., Dahlgaard, J., Fjorback, L.O. and Piet, J., 2015. A systematic review of mechanisms of change in mindfulness-based cognitive therapy in the treatment of recurrent major depressive disorder. Clinical Psychology Review, 37, pp. 26–39.

Viola-Saltzman, M. and Musleh, C., 2016. Traumatic brain injury-induced sleep disorders. *Neuropsychiatric Disease and Treatment*, *12*, p. 339.

Volf, C., Aggestrup, A.S., Svendsen, S.D., Hansen, T.S., Petersen, P.M., Dam-Hansen, C., Knorr, U., Petersen, E.E., Engstrøm, J., Hageman, I. and Jakobsen, J.C., 2020. Dynamic LED light versus static LED light for depressed inpatients: results from a randomized feasibility trial. *Pilot and Feasibility Studies*, *6*(1), pp. 1–9.

von Economo, C., 1930. Sleep as a problem of localization. *Journal of Nerve Mental Disorder*, *71*(3), pp. 249–259.

von Holstein-Rathlou, S., Petersen, N.C. and Nedergaard, M., 2018. Voluntary running enhances glymphatic influx in awake behaving, young mice. *Neuroscience Letters*, *662*, pp. 253–258.

Voshaar, R.C.O., van Balkom, A.J. and Zitman, F.G., 2004. Zolpidem is not superior to temazepam with respect to rebound insomnia: a controlled study. *European neuropsychopharmacology*, *14*(4), pp. 301–306.

Wali, S., Shukr, A., Boudal, A., Alsaiari, A. and Krayem, A., 2015. The effect of vitamin D supplements on the severity of restless legs syndrome. *Sleep and Breathing*, *19*(2), pp. 579–583.

Wali, S., Alsafadi, S., Abaalkhail, B., Ramadan, I., Abulhamail, B., Kousa, M., Alshamrani, R., Faruqui, H., Faruqui, A., Alama, M. and Hamed, M., 2018. The association between vitamin D level and restless legs syndrome: a population-based case-control study. *Journal of Clinical Sleep Medicine*, *14*(4), pp. 557–564.

Walker, M., 2017. *Why we sleep: unlocking the power of sleep and dreams*. Simon and Schuster.

Walker, M.P., 2017. *Why We Sleep: the new science of sleep and dreams*. Penguin.

Walsh, N.P., 2018. *Recommendations to maintain immune health in athletes*. European Journal of Sport Science, 1–12.

Wan, G., Wang, L., Zhang, G., Zhang, J., Lu, Y., Li, J. and Yi, X., 2020. Effects of probiotics combined with early enteral nutrition on endothelin-1 and C-reactive protein levels and prognosis in patients with severe traumatic brain injury. *Journal of International Medical Research*, *48*(3), p. 0300060519888112.

Wassing, R., Lakbila-Kamal, O., Ramautar, J.R., Stoffers, D., Schalkwijk, F. and Van Someren, E.J., 2019. Restless REM sleep impedes overnight amygdala adaptation. *Current Biology*, *29*(14), pp. 2351–2358.

Watson, N.F., Dikmen, S., Machamer, J., Doherty, M. and Temkin, N., 2007. Hypersomnia following traumatic brain injury. *Journal of Clinical Sleep Medicine*, *3*(4), pp. 363–368.

Weber, J.T., 2012. Altered calcium signaling following traumatic brain injury. *Frontiers in Pharmacology*, *3*, p. 60.

Webster, J.B., Bell, K.R., Hussey, J.D., Natale, T.K. and Lakshminarayan, S., 2001. Sleep apnea in adults with traumatic brain injury: a preliminary investigation. *Archives of Physical Medicine and Rehabilitation*, *82*(3), pp. 316–321.

Wheatley, D., 2005. Medicinal plants for insomnia: a review of their pharmacology, efficacy and tolerability. *Journal of Psychopharmacology*, *19*(4), pp. 414–421.

Wheless, J.W., 2008. History of the ketogenic diet. *Epilepsia*, *49*, pp. 3–5.

WHO Report on the Global Tobacco Epidemic, 2008. The MPOWER package. Geneva: World Health Organization.

Wichniak, A., Wierzbicka, A., Wałęcka, M. and Jernajczyk, W., 2017. Effects of antidepressants on sleep. *Current Psychiatry Reports*, *19*(9), pp. 1–7.

Wickwire, E.M., Williams, S.G., Roth, T., Capaldi, V.F., Jaffe, M., Moline, M., Motamedi, G.K., Morgan, G.W., Mysliwiec, V., Germain, A. and Pazdan, R.M., 2016. Sleep, sleep disorders, and mild traumatic brain injury. What we know and what we need to know: findings from a national working group. *Neurotherapeutics*, *13*(2), pp. 403–417.

Wisden, W., Yu, X. and Franks, N.P., 2017. GABA receptors and the pharmacology of sleep. In: *Sleep-wake neurobiology and pharmacology*. Cham: Springer, pp. 279–304.

Wiseman-Hakes, C., Duclos, C., Blais, H., Dumont, M., Bernard, F., Desautels, A., Menon, D.K., Gilbert, D., Carrier, J. and Gosselin, N., 2016. Sleep in the acute phase of severe traumatic brain injury: a snapshot of polysomnography. *Neurorehabilitation and Neural Repair*, *30*(8), pp. 713–721.

Witcher, K.G., Eiferman, D.S. and Godbout, J.P., 2015. Priming the inflammatory pump of the CNS after traumatic brain injury. *Trends in Neurosciences*, *38*(10), pp. 609–620.

Wolf, J.A. and Koch, P.F., 2016. Disruption of network synchrony and cognitive dysfunction after traumatic brain injury. *Frontiers in Systems Neuroscience*, *10*, p. 43.

Wolfe, C.D., Crichton, S.L., Heuschmann, P.U., McKevitt, C.J., Toschke, A.M., Grieve, A.P. and Rudd, A.G., 2011. Estimates of outcomes up to ten years after stroke: analysis from the prospective South London Stroke Register. *PLoS Med*, *8*(5), p. e1001033.

Wu, H.M., Huang, S.C., Hattori, N., Glenn, T.C., Vespa, P.M., Yu, C.L., Hovda, D.A., Phelps, M.E. and Bergsneider, M., 2004. Selective metabolic reduction in gray matter acutely following human traumatic brain injury. *Journal of Neurotrauma*, *21*(2), pp. 149–161.

Wu, T.C., Wilde, E.A., Bigler, E.D., Li, X., Merkley, T.L., Yallampalli, R., McCauley, S.R., Schnelle, K.P., Vasquez, A.C., Chu, Z. and Hanten, G., 2010. Longitudinal changes in the corpus callosum following pediatric traumatic brain injury. *Developmental neuroscience*, *32*(5-6), pp. 361–373.

Xi, M.C., Morales, F.R. and Chase, M.H., 1999. Evidence that wakefulness and REM sleep are controlled by a GABAergic pontine mechanism. *Journal of Neurophysiology*, *82*(4), pp. 2015–2019.

Xu, M., Chung, S., Zhang, S. *et al.* Basal forebrain circuit for sleep-wake control. *Nat Neurosci* **18**, 1641–1647 (2015).

Xu, K., Gao, X., Xia, G., Chen, M., Zeng, N., Wang, S., You, C., Tian, X., Di, H., Tang, W. and Li, P., 2021. Rapid gut dysbiosis induced by stroke exacerbates brain infarction in turn. *Gut*, *70*, pp. 1486–1994.

Xu, X., Wang, L., Chen, L., Su, T., Zhang, Y., Wang, T., Ma, W., Yang, F., Zhai, W., Xie, Y. and Li, D., 2016. Effects of chronic sleep deprivation on bone mass and bone metabolism in rats. *Journal of Orthopaedic Surgery and Research*, *11*(1), pp. 1–9.

Yan, C. and Zhang, B., 2019. Clinical significance of detecting serum melatonin and SBDPs in brain injury in preterm infants. *Pediatrics & Neonatology*, *60*(4), pp. 435–440.

Yang, J., Wang, K., Hu, T., Wang, G., Wang, W. and Zhang, J., 2021. Vitamin D3 supplement attenuates blood–brain barrier disruption and cognitive impairments in a rat model of traumatic brain injury. *NeuroMolecular Medicine*, pp. 1–9. https://doi.org/10.1007/s12017-021-08649-z

Yap, J.Y., Keatch, C., Lambert, E., Woods, W., Stoddart, P.R. and Kameneva, T., 2020. Critical review of transcutaneous vagus nerve stimulation: challenges in translation to clinical practice. *Frontiers in Neuroscience*, *14*. https://doi.org/10.33 89/fnins.2020.00284e

Yarlagadda, K., Ma, N. and Doré, S., 2020. Vitamin D and stroke: effects on incidence, severity, and outcome and the potential benefits of supplementation. *Frontiers in Neurology*, *11*. https://doi.org/10.3389/fneur.2020.00384e

Yoshimura, A. and Ito, M., 2020. Resolution of inflammation and repair after ischemic brain injury. *Neuroimmunology and Neuroinflammation*, *7*(3), pp. 264–276.

Yoshimura A. and Minako, I., 2020. Resolution of inflammation and repair after ischemic brain injury. *Neuroimmunology and Neuroinflammation*, *7*, pp. 264–276.

Young, T., Peppard, P.E. and Gottlieb, D.J., 2002. Epidemiology of obstructive sleep apnea: a population health perspective. *American Journal of Respiratory and Critical Care Medicine*, *165*(9), pp. 1217–1239.

Yurcheshen, M., Seehuus, M. and Pigeon, W., 2015. Updates on nutraceutical sleep therapeutics and investigational research. *Evidence-Based Complementary and Alternative Medicine*, *2015*, 105256. https://doi.org/10.1155/2015/105256e

Zeitzer, J.M., Hon, F., Whyte, J., Monden, K.R., Bogner, J., Dahdah, M., Wittine, L., Bell, K.R. and Nakase-Richardson, R., 2020. Coherence between sleep detection by actigraphy and polysomnography in a multi-center, inpatient cohort of individuals with traumatic brain injury. *PM&R*, *12*(12), pp. 1205–1213.

Zhang, Yan-Ping et al., 2017. Effects of DHA supplementation on hippocampal volume and cognitive function in older adults with mild cognitive impairment: a 12-month randomized, double-blind, placebo-controlled trial. *Journal of Alzheimer's Disease*, *55*(2), pp. 497–507.

Zhang, W.T. and Wang, Y.F., 2017. Efficacy of methylphenidate for the treatment of mental sequelae after traumatic brain injury. *Medicine*, *96*(25), e6960. doi: 10.1 097/MD.0000000000006960

Zhang, Z., Zhang, Z., Lu, H., Yang, Q., Wu, H. and Wang, J., 2017. Microglial polarization and inflammatory mediators after intracerebral hemorrhage. *Molecular Neurobiology*, *54*(3), pp. 1874–1886.

Zhao, B., Li, L., Jiao, Y., Luo, M., Xu, K., Hong, Y., Cao, J.D., Zhang, Y., Fang, J.L. and Rong, P.J., 2019. Transcutaneous auricular vagus nerve stimulation in treating post-stroke insomnia monitored by resting-state fMRI: the first case report. *Brain Stimulation: Basic, Translational, and Clinical Research in Neuromodulation*, *12*(3), pp. 824–826.

Zhu, F., Guo, R., Wang, W., Ju, Y., Wang, Q., Ma, Q., Sun, Q., Fan, Y., Xie, Y., Yang, Z. and Jie, Z., 2020. Transplantation of microbiota from drug-free patients with schizophrenia causes schizophrenia-like abnormal behaviors and dysregulated kynurenine metabolism in mice. *Molecular Psychiatry*, *25*(11), pp. 2905–2918.

Zindler, E. and Zipp, F., 2010. Neuronal injury in chronic CNS inflammation. *Best Practice & Research Clinical AnaestHesiology*, *24*(4), pp. 551–562.

Index

For Product Safety Concerns and Information please contact our EU
representative GPSR@taylorandfrancis.com
Taylor & Francis Verlag GmbH, Kaufingerstraße 24, 80331 München, Germany